高等教育规划教材

卓越工程师教育培养计划系列教材

化工原理实验及虚拟仿真（双语）

Chemical Engineering Principle Experiments and Virtual Simulation（Bilingual）

叶向群　单　岩　主编

Xiangqun Ye　Yan Shan　Editor-in-chief

化学工业出版社

北京·

《化工原理实验及虚拟仿真（双语）》是根据大学化工及其相关专业化工原理实验教学的要求，在原有化工原理实验讲义的基础上编写完成的。其内容包括绪论，化工原理实验研究方法，实验数据的处理方法，化工原理虚拟仿真实验概述，以及对流体力学综合实验——管内流动阻力及离心泵特性曲线测定实验、对流传热系数的测定实验、恒压及真空过滤实验、填料塔吸收过程实验、筛板塔精馏实验、转盘及脉冲萃取实验、洞道式干燥特性曲线测定实验的实验原理、实验装置、操作步骤、虚拟仿真等方面进行了详细介绍。

　　随着现代教育技术的飞速发展，依托于虚拟现实、多媒体、人机交互、数据库和网络通信等技术，构建高度仿真的虚拟实验环境和实验对象，进行虚拟仿真实验教学已经成为实验教学的有效手段。针对教学的需要，本书配套提供了全新的立体教学资源库，内容丰富，形式多样，还配套提供教学工具，使教学效率显著提高。

　　《化工原理实验及虚拟仿真（双语）》可以作为本科院校化工原理实验、化工基础实验等课程的教材，并可供化工、化学等专业的技术人员参考。采用中英文双语编写，对内可用于双语实验教学，对外可用于国际交流生的实验教学。

图书在版编目（CIP）数据

化工原理实验及虚拟仿真（双语）/叶向群，单岩主编. —北京：化学工业出版社，2017.8（2024.8重印）
高等教育规划教材　卓越工程师教育培养计划系列教材
ISBN 978-7-122-30025-6

Ⅰ.①化… Ⅱ.①叶…②单… Ⅲ.①化工原理-实验-高等学校-教材-汉、英 Ⅳ.①TQ02-33

中国版本图书馆 CIP 数据核字（2017）第 146314 号

责任编辑：杜进祥　　　　　　　　　　文字编辑：丁建华　任睿婷
责任校对：王　静　　　　　　　　　　装帧设计：关　飞

出版发行：化学工业出版社（北京市东城区青年湖南街 13 号　邮政编码 100011）
印　　装：北京机工印刷厂有限公司
787mm×1092mm　1/16　印张 14　字数 346 千字　2024 年 8 月北京第 1 版第 2 次印刷

购书咨询：010-64518888（传真：010-64519686）　　售后服务：010-64518899
网　　址：http://www.cip.com.cn
凡购买本书，如有缺损质量问题，本社销售中心负责调换。

定　价：36.00 元　　　　　　　　　　　　　　　　　　　　版权所有　违者必究

前言

化工原理实验作为综合类大学化工类专业及其相关专业重要的专业基础课,是与工程实践相结合的必要条件,也是培养学生工程技术知识的一项重要实践。

化工原理实验作为一门重视实践操作、强调理论积累的历史久远的课程,一直采用了理实一体的授课方法。但由于很多实验设备价格昂贵,实验操作复杂,并具有污染性、高危险性、高能耗、高成本、大空间等特点,为实验教学工作的开展增加了难度。

随着科技的发展,依托于虚拟现实、多媒体、人机交互、数据库和网络通信等技术,构建高度仿真的虚拟实验环境和实验对象并开展教学已经成为实验教学的趋势。在化工原理实验课程中引入虚拟仿真实验,能够显著地提高教学的效率与效果:

① 把现实中的实验操作搬到计算机上进行仿真训练,不仅起到很好的课前预习效果,而且还可大幅降低化工实验操作的危险性并节省成本,同时虚拟仿真实验软件还增加了实验操作的趣味性,达到了寓学于乐的目的;

② 虚拟仿真实验能打破时空限制,为随时随地开展实验教学创造条件;

③ 为教师提供优化的教学环境,并为学生提供个性化学习平台。

本书作为化工原理实验教材,注重理论与实践的结合、实验能力和素质的培养与训练。教师可要求学生通过虚拟仿真实验后自主完成实际实验操作,如根据实验内容自行设计完整的实验数据记录表,根据实验步骤提纲细化实验操作,通过查阅手册获取实验相关数据(如物性参数)、参照使用说明操作常规仪器,等等。适应现代教育技术和学科交叉综合的趋势,强调理论知识、虚拟仿真练习、自主实训操作相结合的"理虚实"一体化教学思路,从而改变以往教师手把手地教、学生拿着表格填数据的传统实验教学模式。

本书由叶向群和单岩主编,杨国成、金伟光、窦梅、南碎飞等参与编写,万文静和钮曹萍参与校核工作。由于编写时间仓促,编者的学识和经验有限,书中必然会存在需要进一步改进和提高的地方,殷切希望广大读者和同行批评指正,使本书日臻完善。

本书的编写过程中,得到浙大旭日科技有限公司在化工原理实验教学软件开发上的大力支持,作为首批国家级化工类虚拟实验中心建设的基础,校企联合配套开发了化工原理虚拟实验室,本书配套提供教学资源库及其云平台——学呗课堂(www.walkclass.com),读者可扫一扫本书封底的二维码下载学呗课堂APP,注册后再扫一扫课程二维码即可在手机端快速获得相应的教学资源。

浙江中控教仪设备有限公司提供了部分实验教学装置,在此表示衷心的感谢。

浙江大学Satmon John博士对本书的英文内容作了校核,在此也深表感谢。

最后,谨向所有为本书提供大力支持的有关学校、企业和领导,以及在组织、撰写、研讨、修改、审定、打印、校对等工作中做出贡献的同志表示由衷的感谢。

编 者

2017年5月1日

目录

绪论 ………………………………… 1
  一、化工原理实验的重要性及其目的 ……………………………… 1
  二、化工原理实验的特点 ……… 1
  三、化工原理实验教学内容与教学方法 ……………………………… 1
  四、本课程化工原理虚拟仿真实验的特点 ………………………… 2
  五、新型的化工原理实验教学方法 … 2

第一章 化工原理实验研究方法 …… 4
  一、直接实验法 ………………… 4
  二、量纲分析法 ………………… 4
  三、数学模型法 ………………… 4
  四、冷模实验法 ………………… 5

第二章 实验数据的处理方法 ……… 6
  一、实验数据误差分析 ………… 6
  二、实验数据处理 ……………… 15

第三章 化工原理虚拟仿真实验概述 …………………………… 22
  一、软件运行环境 ……………… 22
  二、化工原理实验虚拟仿真系统启动 ……………………………… 22
  三、化工原理实验虚拟仿真菜单功能 ……………………………… 23

第四章 流体力学综合实验——管内流动阻力测定实验 ……… 29
  一、实验目的和要求 …………… 29
  二、实验原理 …………………… 29
  三、实验装置和流程 …………… 30
  四、虚拟仿真实验操作步骤 …… 32
  五、实验方法及步骤 …………… 34
  六、实验报告 …………………… 34
  七、思考题 ……………………… 35

第五章 流体力学综合实验——离心泵特性曲线测定实验 …… 36
  一、实验目的和要求 …………… 36
  二、实验原理 …………………… 36
  三、实验装置和流程 …………… 37
  四、虚拟仿真实验操作步骤 …… 38
  五、实验方法及步骤 …………… 39
  六、实验报告 …………………… 39
  七、思考题 ……………………… 39

第六章 对流传热系数的测定实验 … 40
  一、实验目的和要求 …………… 40
  二、实验原理 …………………… 40
  三、实验装置和流程 …………… 43
  四、虚拟仿真实验操作步骤 …… 45
  五、实验方法及步骤 …………… 47
  六、实验报告 …………………… 48
  七、思考题 ……………………… 48

第七章 过滤实验——恒压过滤、真空过滤 …………………… 49
  一、实验目的和要求 …………… 49
  二、实验原理 …………………… 49
  三、实验装置和流程 …………… 51
  四、虚拟仿真实验操作步骤 …… 53
  五、实验方法及步骤 …………… 57
  六、实验报告 …………………… 58
  七、思考题 ……………………… 58

第八章 填料塔吸收过程实验 ……… 59
  一、实验目的和要求 …………… 59

二、实验原理 …………………… 59
　　三、实验装置和流程 …………… 62
　　四、虚拟仿真实验操作步骤 …… 63
　　五、实验方法及步骤 …………… 66
　　六、实验报告 …………………… 66
　　七、思考题 ……………………… 67

第九章　筛板塔精馏实验 ………… 68
　　一、实验目的和要求 …………… 68
　　二、实验原理 …………………… 68
　　三、实验装置和流程 …………… 70
　　四、虚拟仿真实验操作步骤 …… 72
　　五、实验方法及步骤 …………… 74
　　六、实验报告 …………………… 75
　　七、思考题 ……………………… 75

第十章　萃取实验——转盘萃取、脉冲萃取 ……………………… 76
　　一、实验目的和要求 …………… 76
　　二、实验原理 …………………… 76
　　三、实验装置和流程 …………… 80
　　四、虚拟仿真实验操作步骤 …… 82
　　五、实验方法及步骤 …………… 87
　　六、实验报告 …………………… 88
　　七、思考题 ……………………… 88

第十一章　洞道式干燥特性曲线测定实验 ……………………… 89
　　一、实验目的和要求 …………… 89
　　二、实验原理 …………………… 89
　　三、实验装置和流程 …………… 91
　　四、虚拟仿真实验操作步骤 …… 92
　　五、实验方法及步骤 …………… 95
　　六、实验报告 …………………… 95
　　七、思考题 ……………………… 95

Chemical Engineering Principle Experiments and Virtual Simulation (Bilingual) ……… 96

Abstract ……………………………… 97

Preface ……………………………… 98

Introduction ………………………… 100
　Ⅰ　Importance and purpose of chemical engineering principle experiment …… 100
　Ⅱ　Characteristics of chemical engineering principle experiment …… 100
　Ⅲ　Teaching contents and methods of chemical engineering principle experiment …………… 100
　Ⅳ　Characteristics of virtual simulation chemical engineering experiment in this course ………………… 101
　Ⅴ　New teaching method of chemical engineering principle experiment …… 102

Chapter 1　Research Methods of Chemical Engineering Principle Experiment ……………… 105
　Ⅰ　Direct experimental method ……… 105
　Ⅱ　Dimensional analysis method …… 105
　Ⅲ　Mathematical model method …… 106
　Ⅳ　Cold model experimental method … 106

Chapter 2　Processing of Experimental Data ……………………… 107
　Ⅰ　Error analysis of experimental data ……………………………… 107
　Ⅱ　Processing of experimental data ……………………………… 119

Chapter 3　Overview of Virtual Simulation Chemical Engineering Principle Experiment ……………… 128
　Ⅰ　Software operating environment …… 128
　Ⅱ　Virtual simulation system of chemical engineering principle experiment startup ………… 128
　Ⅲ　Menu function of virtual simulation chemical engineering principle experiment ……………………… 129

Chapter 4　Comprehensive Fluid Mechanic Experiment—Determination of Flow Resistance in Pipe ………… 136

- I　Experimental purposes and requirements ……………… 136
- II　Experiment principle ……………… 136
- III　Experimental apparatus and process ……………… 138
- IV　Operation steps of virtual simulation experiment ……………… 139
- V　Experimental method and procedure ……………… 142
- VI　Experimental report ……………… 143
- VII　Questions ……………… 143

Chapter 5　Comprehensive Fluid Mechanic Experiment—Determination of Centrifugal Pump Characteristic Curve ……………… 144

- I　Experimental purposes and requirements ……………… 144
- II　Experiment principle ……………… 144
- III　Experimental apparatus and process ……………… 146
- IV　Operation steps of virtual simulation experiment ……………… 146
- V　Experimental method and procedure ……………… 147
- VI　Experimental report ……………… 148
- VII　Questions ……………… 148

Chapter 6　Determination Experiment of Convective Heat Transfer Coefficient ……………… 149

- I　Experimental purposes and requirements ……………… 149
- II　Experiment principle ……………… 149
- III　Experimental apparatus and process ……………… 153
- IV　Operation steps of virtual simulation experiment ……………… 155
- V　Experimental method and procedure ……………… 158
- VI　Experimental report ……………… 159
- VII　Questions ……………… 159

Chapter 7　Filtration Experiment ……… 160

- I　Experimental purposes and requirements ……………… 160
- II　Experiment principle ……………… 160
- III　Experimental apparatus and process ……………… 162
- IV　Operation steps of virtual simulation experiment ……………… 164
- V　Experimental method and procedure ……………… 169
- VI　Experimental report ……………… 170
- VII　Questions ……………… 171

Chapter 8　Packed Column Absorption Experiment ……………… 172

- I　Experimental purposes and requirements ……………… 172
- II　Experiment principle ……………… 172
- III　Experimental apparatus and process ……………… 175
- IV　Operation steps of virtual simulation experiment ……………… 176
- V　Experimental method and procedure ……………… 180
- VI　Experimental report ……………… 181
- VII　Questions ……………… 181

Chapter 9　Sieve-plate Column Distillation Experiment ……… 183

- I　Experimental purposes and requirements ……………… 183
- II　Experiment principle ……………… 183
- III　Experimental apparatus and process ……………… 186
- IV　Operation steps of virtual simulation experiment ……………… 186
- V　Experimental method and procedure ……………… 191
- VI　Experimental report ……………… 192
- VII　Questions ……………… 192

Chapter 10　Extraction Experiment ……… 193

- I　Experimental purposes and requirements ……………… 193

- II Experiment principle ········· 193
- III Experimental apparatus and process ········· 198
- IV Operation steps of virtual simulation experiment ········· 200
- V Experimental method and procedure ········· 205
- VI Experimental report ········· 207
- VII Questions ········· 207

Chapter 11 Experiment for the Determination of Drying Characteristic Curve in Tunnel Dryer ········· 208

- I Experimental purposes and requirements ········· 208
- II Experiment principle ········· 208
- III Experimental apparatus and process ········· 210
- IV Operation steps of virtual simulation experiment ········· 212
- V Experimental method and procedure ········· 215
- VI Experimental report ········· 215
- VII Questions ········· 215

参考文献 ········· 216

绪　　论

一、化工原理实验的重要性及其目的

化工原理实验是化工类及其相关专业一门实践性很强的课程。不同于化学实验，化工原理实验是研究物料在工程规模条件下，发生物理或化学状态变化的工业过程及这类工业过程所用装置的设计和操作的一门技术课程。一个化工原理实验就是一个单元操作，实际化工生产过程就是由不同类型的单元操作构成。学生在化工原理实验课的学习过程中会遇到大量的工程实际问题，对于理工科学生来讲可以从实验过程中更实际、更有效地学到更多工程方面的原理及测试手段，可以发现复杂的工艺过程与数学模型之间的关系，也可以认识到对于一个看似复杂的过程，可以用最基本的原理来解释和描述。

化工原理实验不仅验证化工原理的基本理论、加深对课堂教学内容的理解，更为重要的还在于对未来的科技工作者进行实验方法、实验技能的基本训练，培养独立组织和完成实验的能力，为将来从事科学研究和解决工程实际问题打好基础。

二、化工原理实验的特点

化工原理实验变量多，涉及的物料千变万化，设备大小悬殊，面对的是复杂的实际问题和工程问题，对象不同，实验研究方法必然不同，本课程内容强调实践性和工程观念，并将能力和素质培养贯穿于实验课的全部过程。围绕化工原理最基本的理论，培养学生掌握实验研究方法，训练其独立思考、综合分析问题和解决问题的能力。

三、化工原理实验教学内容与教学方法

化工原理实验教学主要包括：实验基础知识教学和典型的化工单元操作实验。

实验基础知识教学部分主要讲述化工原理实验教学的目的和要求；实验装置和流程；实验原理；实验方法和步骤；实验报告等相关知识。

化工单元操作实验部分主要有：管内力学综合实验——管内流动阻力测定实验、离心泵特性曲线测定实验；对流传热系数的测定实验；过滤实验——恒压过滤、真空过滤实验；填料塔吸收实验；筛板塔精馏操作及效率测定实验；萃取塔（转盘塔/脉冲塔）操作实验；洞

道式干燥特性曲线测定实验。

化工原理实验教学一直采用了实验理论讲解和实验过程相结合的传统授课方法。传统教学方法具有明显的缺点：

① 很多实验设备价格昂贵，实验操作复杂，并往往具有污染性、高危险性（如有毒有害原料或试剂、高温高压或高速旋转等）、高能耗、高成本、大空间等特点，为实践教学工作的开展增加了难度。

② 通常是在教师完成理论授课后直接为学生安排实验，但理论与真实实验之间往往跨度很大，在一些操作复杂、设备不足、学生人数又多的情况下，学生往往会因为对实验没有充分的理解和准备而在现场实验中手足无措，导致成学习效率低下、浪费时间和资源，甚至会增加学生的心理负担。

③ 由于实验设备及场所有限制，往往需要把学生分成很多批次，很多学生的实验时间与理论授课时间间隔很长，不仅大幅度增加了教师的工作量，而且实验还没有起到应有的效果。

四、本课程化工原理虚拟仿真实验的特点

本课程虚拟仿真实验教学是依托虚拟现实、多媒体、人机交互、数据库和网络通信等技术，构建高度仿真的虚拟实验环境和实验对象并开展教学。化工原理虚拟仿真实验教学注重理工结合、教学与科研互动、校内与企业的衔接，并强调以学生为本，通过虚实串行、虚实并行、以虚替实和远程控制的多样性、开放式实验教学训练，既保证了学生的实验动手能力的训练，又充分利用虚拟仿真实验的特点，强化学生对实验过程的深入认识和对知识点的深入理解，使学生能从微观实验看到宏观效果、从教学实验中看到科研内涵、从校内看到校外生产实际，有效提升实验教学的效果，并实现化工类实验教学的绿色化，确保化工实验过程的安全性。

在化工原理实验课程中引入虚拟仿真实验手段，具有很显著的优点：

① 把现实中的实验操作搬到计算机上进行仿真训练，不仅可以起到很好的课前预习效果，还能降低化工实验操作的危险性，节省成本，同时虚拟仿真实验还增加了实验操作的趣味性，达到了寓学于乐的目的。

② 虚拟仿真实验不仅能打破时空限制，为随时随地开展实验教学创造便利条件，而且可以用虚拟仿真实验代替真实实验，避免或大幅度降低真实化工类实验所带来的各种危险。

③ 虚拟仿真实验为教师的教学活动提供了优化的教学环境。虚拟仿真实验中的教学资源内容丰富、形式多样、专业针对性强，为教师的备课提供了一个数字化平台。教师可以根据教学目标和学生的需要，设计更好的教学过程，达到最优化的课堂教学效果。

④ 虚拟仿真实验还为学生营造了数字化的学习环境，全面支持学生的自主化学习、研究性学习等多种学习模式。学生在数字化学习环境中，充分发挥自主性，进行探究、协作、自主和创造性学习，从而提高自身的创新精神、协作精神和实践能力。

五、新型的化工原理实验教学方法

1. "理虚实"一体化教学方法

"理虚实"一体化教学新模式是在理论授课后，引入虚拟仿真实验环节。将虚拟仿真实验作为实验前的准备和演练，为学生搭建了从理论到实际的桥梁，易学易用，是本课程"理

虚实"一体化教学思路的亮点所在。

① 在课前，向学生提供理论学习资源、虚拟仿真软件及学习任务。理论资源包括实验指导书、实验注意事项、实验操作视频等。通过个性化学习，让学生掌握相关理论知识，并了解相应实验设备的操作。

② 课堂中，首先进行理论测验和虚拟仿真软件的虚拟实验考核，以了解学生对理论知识和实验操作的掌握程度。只有通过理论测验，特别是虚拟实验考核的学生才能进行真实的实验。

③ 真实实验。观察实验现象并记录实验数据。由于虚拟实验与真实实验的操作步骤完全一致，不仅可以很好地解决真实实验的安全性问题，更可以让学生将精力集中于对实验及理论的深入理解。

如上述，"理虚实"一体化教学模式能否成功实施，关键在于如何保证学生能在课外进行理论学习和虚拟仿真实训，以及虚拟仿真实验的逼真程度。为此，通过校企联合开发了化工原理虚拟实验室以及云平台。

云平台的作用是实现资源共享、在线学习、教学管理等，实现资源的高效传播、任务发布、在线测验、在线使用虚拟仿真软件等。基于云平台的教学，不仅是对传统的教学方式的一种突破，也是互联网时代随时随地、人人都可参与的资源共享精神的一种表达。

所开发的虚拟仿真实验软件则同时具备教、练、考三个功能，"教"部分是通过视频演示实验操作流程，让学生初步了解正确的实验操作步骤。"练"部分是允许学生反复练习实验的流程，加强巩固。"考"的部分是考查学生对实验掌握程度的一个测验，它会自动记录学生在这个过程中产生的操作信息并对操作步骤的正确与否做出评价，有利于学生了解自己的知识掌握情况并进行查漏补缺；它还能自动记录实验数据，生成相应的 Excel 表格，清晰、整洁、准确，便于学生收集实验数据，完成实验分析和报告。

虚拟仿真软件还实现与云平台的无缝集成，即具备从 Web 页面启动软件、实验数据自动记录、虚拟实验操作在线考试、实验报告上传等快捷易用的功能，让虚拟仿真实验随时随地都能进行，让教师随时随地了解学生学习情况。

基于"理虚实"一体化的化工原理新型实验教学模式，极大地缓解了实验室资源的压力，不仅有助于节约教学成本，也大大提高了教学效果。

2. "理虚实"一体化教学实践

以"理虚实"一体化教学模式开展化工原理实验，首先是将教学资源上传到云平台，并通过云平台下达学习任务和设定学习目标，学生收到通知后再根据学习任务和目标，使用相关理论资源并进行虚拟仿真化工实验。学生完成虚拟仿真化工实验软件的练习和考试后，软件会自动保存实验操作过程和实验数据，并给相关的操作步骤评分，最终的数据将自动提交到云平台。教师登录云平台管理并统计学生的预习、实验操作情况。学生通过考核后再到真实的实验室进行实验，实验报告完成后，也提交到云平台，由教师进行在线批阅。

"理虚实"一体化教学模式深受学生欢迎。虚拟仿真化工软件能很好地帮助学生做好充足的课前准备、完成自主预习，短视频教学中展示的实验操作能帮助学生在实验前做好预习工作、熟悉实验设备；它不仅是很好的自学工具，甚至还可以为对实验原理的理解提供灵感和洞见，解决一些疑惑；加上其逼真的画面和三维漫游用户体验，是其在被访学生群体中受欢迎的主要原因。"理虚实"一体化教学可以使学生在无人指导的情况下完成课前预习，为没有实验条件的学生提供替代性的教学资源。

第一章
化工原理实验研究方法

化工原理是一门工程学科，它要解决的不单是过程的基本规律，而且面临着真实、复杂的生产问题——特定的物料在特定的设备中进行特定的过程。实际问题的复杂性不完全在于过程本身，而首先在于化工设备的复杂的几何形状和千变万化的物性。对于化学工程学科，除了生产经验总结以外，实验研究是学科建立和发展的重要基础。在长期的经验总结和实验研究的基础上，化工原理实验逐步形成的研究方法主要有直接实验法、量纲分析法、数学模型法和冷模实验法。

一、直接实验法

直接实验法即对被研究对象进行直接的实验，控制或模拟某些客观条件，以获取其相关的参数及规律。这是一种解决工程实际问题的最基本的方法，直接有效，所得到的结果也较为可靠，但这个方法也有较大的局限性。直接实验法得出的只是个别参数之间关系的规律，不能反映对象的全部本质，这些实验结果只能用到特定的实验条件和实验设备上，或推广到实验条件完全相同的现象。另外实验工作量大，耗时耗力，有时需要较高的投资。

二、量纲分析法

化工原理实验面对的是多变量影响的工程问题，需要借助于实验研究方法建立经验关系式，来导出理论公式。而研究多变量影响过程的规律，往往采用固定其他变量，依次改变其中某一个变量的网格法进行实验。如果变量数为 m 个，每个变量改变条件数为 n 次，那么所需要实验的次数为 $n \times m$ 次，涉及变量多，所需实验次数就会剧增，实验工作量必然很大。为了减少实验的工作量并使得到的实验结果具有一定的普遍性，采用量纲分析法这个化工原理广泛使用的实验研究方法来解决这类问题。

三、数学模型法

数学模型是用符号、函数关系将评价目标和内容系统规定下来，并把互相间的变化关系通过数学公式表达出来。所表达的内容可以是定量的，也可以是定性的，但必须以定量的方式体现出来。因此，数学模型法的操作方式偏向于定量形式。基本特征：评价问题抽象化和仿真化；各参数是由与评价对象有关的因素构成的；要表明各有关因素之间的关系。

数学模型法是在对研究的问题有充分认识的基础上，按以下主要步骤进行工作：

① 将复杂问题做合理又不过于失真的简化，提出一个近似实际过程又易于用数字方程式描述的物理模型；

② 对所得到的物理模型进行数学描述建立数学模型，然后确定该方程的初始条件和边界条件，求解方程；

③ 通过实验对数学模型的合理性进行检验并测定模型参数。

四、冷模实验法

冷模实验主要用于流动状态、传递过程等物理过程模拟研究，通过模拟实验结果去分析、推测实际过程。例如利用空气和水并加入示踪剂可进行气液传质的实验研究，为气液传质设备的设计和改造提供参数；利用空气和砂进行流态化的实验研究，为流化床反应器设计提供依据。因此利用空气、水、砂等模拟物料替代真实物料在工业装置结构尺寸相似实验装置中研究各种工程因素对过程影响规律的实验称为"冷模实验"。

冷模实验结果可推广应用于其他实际流体，将小尺寸实验设备的实验结果推广应用于大型工业装置，使得实验能够在物料种类上"由此及彼"，在设备尺寸上"由小及大"，直观、经济。用少量实验，结合数学模型法或量纲分析法，可求得各物理量之间的关系，使实验工作量大为减少。可进行在真实条件下不便或不可能进行的类比实验，减少实验的危险性。但是，冷模实验结果必须结合化学反应等特点，进行校正后才可用于工业过程的设计和开发。

第二章 实验数据的处理方法

通过实验测量所得大批原始数据是实验的主要成果,而后需要进行计算将最终的实验结果归纳成经验公式或者以图表的形式表示,以便进行比较分析。但在实验过程中,由于测量仪器、操作方法和人的观察等方面原因,数据总存在一些误差,所以在整理这些原始数据时,首先应对实验数据的可靠性进行客观的评定。

一、实验数据误差分析

误差分析的目的就是评定实验数据的准确性,通过误差分析,认清误差的来源及其影响,并设法排除数据中所包含的无效成分。在实验中注意哪些是影响实验精确度的主要方面,还可以进一步改进实验方案,细心操作,从而提高实验的精确度。因此,对实验误差进行分析和估算,在评判实验结果和设计方案方面具有重要意义。

1. 真值与平均值

真值是待测物理量客观存在的确定值,也称理论值或定义值。通常一个物理量的真值是不知道的,需要去测定它。但严格来讲,由于测量仪器、测定方法、环境、人的观察力、测量的程序等,都不能做到完美无缺,故真值是无法测得的。为了使真值这个名词不致太玄虚,可以这样定义实验科学中的真值:在实验中,测量的次数无限多时,根据误差的分布定律,正负误差的出现概率相等,再经过细致地消除系统误差,将测量值加以平均,可以获得非常接近于真值的数值。但是实际上实验测量的次数总是有限的。用有限测量值求得的平均值只能是近似真值,常用的平均值有下列几种:

(1) 算术平均值

算术平均值是最常见的一种平均值。

设 $x_1, x_2, \cdots, x_n$ 为各次测量值,n 代表测量次数,则算术平均值为

$$\bar{x} = \frac{x_1 + x_2 + \cdots + x_n}{n} = \frac{\sum\limits_{i=1}^{n} x_i}{n} \tag{2-1}$$

(2) 几何平均值

几何平均值是将一组 n 个测量值连乘并开 n 次方求得的平均值。即

$$\bar{x}_n = \sqrt[n]{x_1 x_2 \cdots x_n} \tag{2-2}$$

（3）均方根平均值

$$\bar{x}_s = \sqrt{\frac{x_1^2 + x_2^2 + \cdots + x_n^2}{n}} = \sqrt{\frac{\sum_{i=1}^{n} x_i^2}{n}} \tag{2-3}$$

（4）对数平均值

在化学反应、热量和质量传递中，其分布曲线多具有对数的特性，在这种情况下表征平均值常用对数平均值。

设两个量 x_1、x_2，其对数平均值

$$\bar{x}_m = \frac{x_1 - x_2}{\ln x_1 - \ln x_2} = \frac{x_1 - x_2}{\ln \frac{x_1}{x_2}} \tag{2-4}$$

应指出，变量的对数平均值总小于算术平均值。当 $x_1/x_2 \leqslant 2$ 时，可以用算术平均值代替对数平均值。

当 $x_1/x_2 = 2$，$\bar{x}_m = 1.443$，$\bar{x} = 1.50$，$(\bar{x}_m - \bar{x})/\bar{x}_m = 4.2\%$，即 $x_1/x_2 \leqslant 2$，引起的误差不超过 4.2%。

（5）加权平均值

设对同一物理量用不同方法去测定，或对同一物理量由不同人去测定。即不等精度测量的每个测量值的可靠性

$$w = \frac{w_1 x_1 + w_2 x_2 + \cdots + w_n x_n}{w_1 + w_2 + \cdots + w_n} = \frac{\sum w_i x_i}{\sum w_i} \tag{2-5}$$

式中，w_1，w_2，…，w_n 表示各观察值的对应权数。各观察值的权数一般凭经验确定。

（6）中位值

指将一组观测值按一定大小次序排列时的中间值，若观测次数为偶数，则中位值为正中两个值的平均值。中位值的最大优点是求法简单，而与两端的变化无关。中位值在设计上属于一种次序统计，只有观测值的分布为正常分布时，它才能代表一组观测值的最佳值。

以上介绍各平均值的目的是要从一组测定值中找出最接近真值的那个值。在化工实验和科学研究中，数据的分布较多属于正态分布，所以通常采用算术平均值。

2. 误差的性质及其分类

在任何一种测量中，无论所用仪器多么精密，方法多么完善，实验者多么细心，不同时间所测得的结果不一定完全一致，而有一定的误差和偏差。严格来讲，误差是指实验测量值（包括直接和间接测量值）与真值（客观存在的准确值）之差，偏差是指实验测量值与平均值之差，但习惯上通常不对两者加以区别。根据误差的性质及其产生的原因，可将误差分为系统误差、偶然误差和过失误差三种。

① 系统误差（恒定误差） 在相同的实验条件下，对同一量进行多次测量时，误差的数值大小和正负始终保持不变，或随着实验条件的改变按一定的规律变化的误差，称为系统误差。例如：刻度不准、零点未校准的测量仪器；实验环境的变化，如外界温度、压力、湿度的变化；实验操作者的习惯与偏向等。

由于系统误差是测量误差的重要组成部分，消除和估计系统误差对于提高测量准确度十分重要。一般系统误差是有规律的，其产生的原因往往是可知的，找出原因后，经过精心校

正或检查可以消除。

② 随机误差（偶然误差）　在相同条件下，测量同一物理量时，误差的绝对值时大时小，符号时正时负，没有一定的规律且无法预测，但这种误差完全服从统计规律，对同一物理量做多次测量，随着测量次数的增加，随机误差的算术平均值趋近于零。因此多次测量的算术平均值将接近于真值。

③ 过失误差　由于实验操作人员操作错误或人为失误所产生的误差。这类误差往往表现为与正常值相差很大，在数据整理时应予以剔除。

3. 误差的表示方法

利用任何量具或仪器进行测量时，总存在误差，测量结果不可能准确地等于被测量的真值，而只是它的近似值。测量的质量高低以测量精确度作指标，根据测量误差的大小来估计测量的精确度。测量结果的误差愈小，则认为测量就愈精确。

测量误差分为测量点和测量列（集合）的误差，它们有不同的表示方法。

(1) 测量点的误差表示方法

① 绝对误差 D　测量集合中某次测量值 x 与其真值 A_0 之差的绝对值称为绝对误差。它的表达式为：

$$D = | x - A_0 | \tag{2-6}$$

即

$$x - A_0 = \pm D \quad x - D \leqslant A_0 \leqslant x + D \tag{2-7}$$

由于真值 A_0 一般无法求得，因而上式只有理论意义。常用高一级标准仪器的示值作为实际值 A 以代替真值 A_0。由于高一级标准仪器存在较小的误差，因而 A 不等于 A_0，但总比 x 更接近于 A_0。x 与 A 之差称为仪器的示值绝对误差。记为：

$$d = | x - A | \tag{2-8}$$

与 d 相反的数称为修正值，记为：

$$C = -d \tag{2-9}$$

通过检定，可以由高一级标准仪器给出被检仪器的修正值 C。利用修正值便可以求出该仪器的实际值 A。

② 相对误差　衡量某一测量值的准确程度，一般用相对误差来表示。示值绝对误差 d 与被测量的实际值 A 的百分比值称为实际相对误差。记为：

$$E_r(x) = \frac{d}{A} \times 100\% \tag{2-10}$$

以仪器的示值 x 代替实际值 A 的相对误差称为示值相对误差。记为：

$$E_r(x) = \frac{d}{x} \times 100\% \tag{2-11}$$

一般来说，除了某些理论分析外，用示值相对误差较为适宜。

③ 引用误差　为了计算和划分仪表精确度等级，提出引用误差概念。其定义为仪表示值的绝对误差与量程范围之比。

$$\delta_A = \frac{示值绝对误差}{量程范围} \times 100\% = \frac{d}{X_n} \times 100\% \tag{2-12}$$

式中　d——示值绝对误差；

　　　X_n——标尺上限值－标尺下限值。

(2) 测量列（集合）的误差表示方法

① 范围误差　范围误差是指一组测量中的最高值与最低值之差，以此作为误差变化的范围。使用中常应用误差系数的概念。

$$K_1 = \frac{L}{\bar{x}} \tag{2-13}$$

式中　K_1——最大误差系数；
　　　L——范围误差（一组测量中的最高值与最低值之差）；
　　　$\bar{x}$——算术平均值。

范围误差最大缺点是 K_1 只取决于两极端值，而与测量次数无关。

② 算术平均误差　算术平均误差是各个测量点的误差的平均值。

$$\delta = \frac{\sum\limits_{i=1}^{n} |d_i|}{n} \tag{2-14}$$

式中　n——测量次数；
　　　d_i——测量值与平均值的偏差，$d_i = x_i - \bar{x}$。

算术平均误差是表示误差的较好方法，它的缺点是无法表示出各次测量间彼此符合的情况。

③ 标准误差　标准误差亦称为均方根误差。其定义为

$$\sigma = \sqrt{\frac{\sum\limits_{i=1}^{n} d_i^2}{n}} \tag{2-15}$$

标准误差对一组测量中的较大误差或较小误差感觉比较灵敏，是表示精确度的较好方法。

式 (2-15) 适用无限次测量的场合。实际测量工作中，测量次数是有限的，则改用式 (2-16)

$$\sigma = \sqrt{\frac{\sum\limits_{i=1}^{n} d_i^2}{n-1}} \tag{2-16}$$

标准误差 σ 不是一个具体的误差，它的大小只说明在一定条件下等精度测量集合所属的每一个观测值对其算术平均值的分散程度。如果 σ 的值愈小，则说明每一次测量值对其算术平均值分散度就愈小，测量的精度就愈高，反之精度就愈低。

在化工原理实验中最常用的 U 形管压差计、转子流量计、秒表、量筒、电压表等仪表原则上均取其最小刻度值为最大误差，而取其最小刻度值的一半作为绝对误差计算值。

④ 或然误差　或然误差也称概差，用符号 γ 表示，它的意义为：在一组测量中若不计正负号，误差大于 γ 的观测值与误差小于 γ 的观测值将各占观测次数的 50%。

$$\gamma = 0.6745\sigma \tag{2-17}$$

或然误差近年来已逐渐被标准误差所替代。

上述的各种误差表示方式中，不论是比较各种测量的精确度或是评定测量结果的质量，均以相对误差和标准误差表示为佳，而在文献中标准误差更常被采用。

4. 实验数据的精密度、正确度和精确度

在实验过程中，往往满足于实验数据的重现性，而忽略了数据测量值的准确程度。绝对真值是不可知的，人们只能订出一些国际标准作为测量仪表准确性的参考标准。国内外文献中所用的名词术语颇为不统一，精密度、正确度、精确度这几个术语的使用一向比较混乱。近来趋于一致的多数意见是：

① 精密度　测量中所测得数值重现性的程度，称为精密度。它反映偶然误差的影响程度，精密度高就表示偶然误差小。

② 正确度　指在规定条件下，测量中所有系统误差的综合，它反映系统误差的大小。

③ 精确度（准确度）　测量值与真值的偏移程度。它反映系统误差和随机误差综合大小的程度。

为了说明精密度与精确度的区别，可用下述打靶子例子来说明。

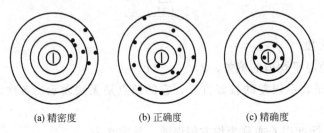

(a) 精密度　　　　(b) 正确度　　　　(c) 精确度

图 2-1　精密度、正确度和精确度的关系

如图 2-1 所示，图 (b) 的系统误差小而随机误差大，即正确度高而精密度低；图 (a) 的系统误差大而随机误差小，即正确度低而精密度高；图 (c) 的系统误差和随机误差都小，表示精确度高。当然在科学测量中没有像靶心那样明确的真值，而是设法去测定这个未知的真值。

对于实验或测量来说，精密度高，正确度不一定高；正确度高，精密度也不一定高。但精确度高，则必须是精密度和正确度都高。

绝对真值是不可知的，随着人类认识运动的推移和发展，可以逐步逼近绝对真值。

5. 仪表的精确度与测量值的误差

(1) 电工仪器等一些仪表的精确度与测量误差

这些仪表的精确度常采用仪表的最大引用误差和精确度等级来表示。仪表的最大引用误差定义为：

$$最大引用误差 = \frac{仪表显示值的绝对误差}{该仪表相应档次量程的绝对值} \times 100\% \quad (2-18)$$

式中，仪表显示值的绝对误差指在规定的正常情况下，被测参数的测量值与被测参数的标准值之差的绝对值的最大值。对于多档仪表，不同档次显示值的绝对误差和量程范围均不相同。

式 (2-18) 表明，若仪表显示值的绝对误差相同，则量程范围愈大，最大引用误差愈小。

我国电工仪表的精确度等级有七种：0.1，0.2，0.5，1.0，1.5，2.5，5.0。如某仪表等级为 2.5 级，则说明此仪表的最大引用误差为 2.5%。

在使用仪表时，如何估算一个测量值的绝对误差和相对误差？

设仪表的精确度等级为 P 级，其最大引用误差为 $P\%$。设仪表的测量范围为 x_n，仪表的测量值为 x_i，则由式（2-18）得该示值的误差为：

$$\text{绝对误差 } D \leqslant x_n \times P\% \tag{2-19}$$

$$\text{相对误差 } E_r(x) = \frac{D}{x_i} \leqslant \frac{x_n}{x_i} \times P\% \tag{2-20}$$

式（2-20）表明：

① 若仪表的精确度等级 P 和测量范围 x_n 已固定，则测量的示值 x_i 愈大，测量的相对误差愈小。

② 选用仪表时，不能盲目地追求仪表的精确度等级，因为测量的相对误差还与 $\frac{x_n}{x_i}$ 有关，应该兼顾仪表的精确度等级和 $\frac{x_n}{x_i}$ 两者。

(2) 天平类仪器的精确度与测量关系

这些仪器的精确度用式（2-21）表示：

$$\text{仪表的精确度} = \frac{\text{名义分度值}}{\text{量程的范围}} \tag{2-21}$$

式中，名义分度值（感量）指测量时读数有把握正确的最小分度单位，即每个最小分度所代表的数值。例如 TG-328A 型天平，其名义分度值为 0.1mg，测量范围为 0~200g，则其

$$\text{精确度} = \frac{0.1}{(200-0) \times 10^3} = 5 \times 10^{-7} \tag{2-22}$$

若仪器的精确度已知，也可用式（2-21）求得其名义分度值。

使用这些仪器时，测量值的误差可用下式来确定：

$$\begin{cases} \text{绝对误差} \leqslant \text{名义分度值} \\ \text{相对误差} \leqslant \dfrac{\text{名义分度值}}{\text{测量值}} \end{cases}$$

(3) 测量值的实际误差

根据仪表的精确度，用上述方法所确定的测量误差，一般总是比测量值的实际误差小得多。这是因为仪器没有调整到理想状态，如不垂直、不水平、零位没有调整好等，会引起误差；仪表的实际工作条件不符合规定的正常工作条件，会引起附加误差；仪器经过长时间使用后，零件发生磨损，装配状态发生变化等，也会引起误差；可能存在有测量者个人的使用习惯和偏向引起的误差；仪表所感受的信号实际上可能并不等于待测的信号；仪表电路可能会受到干扰……

总而言之，测量值实际误差大小的影响因素是很多的。为了获得较准确的测量结果，需要有较好的仪器，也需要有科学的作风和方法，以及扎实的理论知识和实践经验。

6. 有效数字与运算法则

在科学与工程中，测量或计算结果总是以一定位数的数字来表示，不是说一个数值中小数点后面位数越多越准确。实验中从测量仪表上所读数值的位数是有限的，其最后一位数字往往是仪表精度所决定的估计数字。即一般应读到测量仪表最小分度的十分之一。数值准确度大小由有效数字位数来决定。

实验直接测量或计算结果，该用几位数字来表示，是件很重要的事情。学生往往容易产生这两种想法：一个数值中小数点后面位数愈多愈准确，或者计算结果保留位数愈多愈准确。其实这两种想法都是错误的。因为其一，小数点的位置不决定准确度，而与所用单位大小有关。例如，用电位差计测热电偶的电动势记为 $764.9\mu V$ 或记为 $0.7649mV$，准确度是完全相同的；其二，测量仪器只能达到一定精度（或称灵敏度），还以上面的例子来说，这种电位差计精度只能达到 $0.1\mu V$ 或 $0.0001mV$，运算结果的准确度绝不会超过这个仪器所允许的误差范围。

由此可见，测量值或计算结果数值用几位数字来表示，决定于测量仪器的精度。数值准确度大小，由有效数字位数来决定。如上面例子中，数值的精度为 $0.1\mu V$，准确度为四位有效数字。

（1）有效数字的概念

一个数据，其中除了起定位作用的"0"外，其他数都是有效数字。如 0.0037 只有两位有效数字，而 370.0 则有四位有效数字。要注意有效数字不一定都是可靠数字。

① 直接测量数据的有效数字　实验中所测得的数据都只能是近似值。通常测量时，一般可读到仪表最小刻度的后一位数，最后一位数是估计数字，它包含在有效数字内。如二等标准温度计最小刻度为 $0.1℃$，我们可以读到 $0.01℃$，如 $15.16℃$。此时有效数字为 4 位，而可靠数字只有三位，最后一位是不可靠的，称为可疑数字。记录测量数值时只保留一位可疑数字。读数为 $15.2℃$ 时，应记为 $15.20℃$，表明有效数字为 4 位。

为了清楚地表示出数值的准确度，明确读出有效数字位数，常用指数的形式表示，即写成一个小数与相应 10 的整数幂的乘积。这种以 10 的整数幂来记数的方法称为科学记数法。

例如：75200　　有效数字为 4 位时，记为 7.520×10^4
　　　　　　　有效数字为 3 位时，记为 7.52×10^4
　　　　　　　有效数字为 2 位时，记为 7.5×10^4

又如：0.00478　有效数字为 4 位时，记为 4.780×10^{-3}
　　　　　　　有效数字为 3 位时，记为 4.78×10^{-3}
　　　　　　　有效数字为 2 位时，记为 4.8×10^{-3}

测量时取几位有效数字取决于对实验结果精确度的要求及测量仪表本身的精确度。

② 非直接测量值的有效数字　在实验中，除使用上一类有单位的数字外，还会碰到另一类没有单位的常数，如 π、e 等，以及某些因子，如 $\sqrt{2}$ 等。它们的有效数字位数，可以认为是无限的。引用它们时取几位有效数字为好，取决于计算所用的原始数据的有效数字的位数。假设参与计算的原始数据中，位数最多的有效数字是 n 位，则引用上述常数时宜取 $n+2$ 位，目的是避免引用数据介入而造成更大的误差。

在数据计算过程中，为使计算结果精确度尽可能高一些，所有的中间计算结果，均可比原始实验数据中有效数字最多者多取 2 位。但在回归分析计算中，中间结果的有效数字位数越多越好，至少应取 6 位，这样可减弱舍入误差的迅速累积。

表示误差大小的数据，一般宜取 2 位有效数字。

（2）有效数字的运算

通过运算后所得到的结果其准确度不可能超过原始记录数据，所以计算过程中，一个数据的位数保留过多，并不能提高精度，反而浪费时间，位数取得过少，会降低应有的精度。运算中数字位数的取舍是根据有效数字运算规则确定的。

① 在加减计算中，计算结果所保留的小数点后的位数，应与所给各数中小数点后位数最少的相同。例如 13.65，0.0082，1.632 三个数目相加时，应写为

$$13.65 + 0.01 + 1.63 = 15.29$$

相加时，将小数点后面的位数均计入，未知数均代以 x，则得

$$
\begin{array}{r}
1\,3\,.\,6\,5\,x\,x \\
0\,.\,0\,0\,8\,2 \\
+\quad 1\,.\,6\,3\,2\,x \\
\hline
1\,5\,.\,2\,9\,x\,x
\end{array}
$$

可证明 15.29 是最合理的答案。

② 在乘除计算中，其结果的有效数字位数，应与所给各数中有效数字位数最少的相同。如 $1.3048 \times 236 = 307.9328$ 取结果 308（四舍五入）是根据 236 这个数值取舍的。

③ 在乘方、开方计算中，其结果的有效数字位数应与其底的有效数字位数相等。

④ 在对数计算中，所取的对数应与真数有效数字位数相等（不包括定位部分）。

⑤ 计算时，第一位有效数字等于或大于 8 时，有效数字位数可增加一位，例如 8.13 实际上只有三位有效数字，但在计算时可做四位计算。

(3) 数字舍入原则

由于计算或其他原因，实验结果数字位数较多时，需将数字截到所要求的位数，最好采用以下舍入原则：

① 拟舍弃数字的最左一位数字小于 5 时，则舍去，即保留的各位数字不变。

② 拟舍弃数字的最左一位数字大于 5，或者是 5，而其后跟有非 0 的数字时，则进 1，即保留的末位数字加 1。

③ 拟舍弃数字的最左一位数字为 5，而 5 的右边无数字或皆为 0 时，若所保留的末位数字为奇数则进 1，为偶数或 0 则舍去，即"单进双不进"。

上述规则也称数字修约的偶数规则，即"四舍六入五凑偶"规则。

例如：2.8635 取四位有效数字时为 2.864

取三位有效数字时为 2.86

2.8665 取四位有效数字时为 2.866

取三位有效数字时为 2.87

2.866501 取四位有效数字时为 2.867

2.86499 取三位有效数字时为 2.86

7. 疏失误差的舍弃

测量中有时会出现少量过大或过小的数值，这些异常值将对测量结果产生很坏的影响，应从测量结果中舍弃，但随意舍弃这一"坏值"以获得实验结果的一致性是不对的。因为随意地舍弃一些误差大的不属于异常值的测量值，会产生虚假的测量结果。

判断是否属于异常值最简单的方法是三倍标准误差判据。

从概率的理论可知，大于 3σ（均方根误差）的误差的出现概率只有 0.3%，故通常把这一数值称作极限误差，即：

$$\delta_{\text{limit}} = 3\sigma \tag{2-23}$$

如果个别测量的误差超过 3σ，那么就可认为属于疏失误差而将其舍弃。重要的是如何从少数几次测量值中舍弃可疑值的问题。因为测量次数少，概率理论已不适用，而个别失常

测量值对算术平均值影响又很大。

有人曾提出一个简单的判断法,即略去可疑测量值后,计算其余各测量值的平均值及平均误差 δ,然后算出可疑测量值与平均值的偏差 d。如果 $d \geqslant 4\delta$,则此可疑值存在的概率大约只有千分之一。

8. 间接测量中的误差传递

在许多实验和研究中,所需要的结果不是用仪器直接测量得到的,而是要把一些直接测量值代入一定的理论关系式中,通过计算才能求得所需要的结果,即间接测量值。由于直接测量值总有一定的误差,因此它们必然引起间接测量值也有一定的误差,即直接测量误差不可避免地传递到间接测量中去,而产生间接测量误差。

误差传递公式:从数学中知道,当间接测量值(y)与直接测量值(x_1, x_2, …, x_n)有函数关系时,即

$$y = f(x_1, x_2, \cdots, x_n) \tag{2-24}$$

则其微分式为:

$$\mathrm{d}y = \frac{\partial y}{\partial x_1}\mathrm{d}x_1 + \frac{\partial y}{\partial x_2}\mathrm{d}x_2 + \cdots + \frac{\partial y}{\partial x_n}\mathrm{d}x_n \tag{2-25}$$

$$\frac{\mathrm{d}y}{y} = \frac{1}{f(x_1, x_2, \cdots, x_n)}\left[\frac{\partial y}{\partial x_1}\mathrm{d}x_1 + \frac{\partial y}{\partial x_2}\mathrm{d}x_2 + \cdots + \frac{\partial y}{\partial x_n}\mathrm{d}x_n\right] \tag{2-26}$$

根据此二式,当直接测量值的误差(Δx_1, Δx_2, …, Δx_n)很小,并且考虑到最不利的情况,应是误差累计和取绝对值,则可求间接测量值的误差(Δy 和 $\frac{\Delta y}{y}$)为:

$$\Delta y = \left|\frac{\partial y}{\partial x_1}\right| \cdot |\mathrm{d}x_1| + \left|\frac{\partial y}{\partial x_2}\right| \cdot |\mathrm{d}x_2| + \cdots + \left|\frac{\partial y}{\partial x_n}\right| \cdot |\mathrm{d}x_n| \tag{2-27}$$

$$E_r(y) = \frac{\Delta y}{y} = \frac{1}{f(x_1, x_2, \cdots, x_n)}\left[\left|\frac{\partial y}{\partial x_1}\right| \cdot |\mathrm{d}x_1| + \left|\frac{\partial y}{\partial x_2}\right| \cdot |\mathrm{d}x_2| + \cdots + \left|\frac{\partial y}{\partial x_n}\right| \cdot |\mathrm{d}x_n|\right] \tag{2-28}$$

这两个式子就是由直接测量误差计算间接测量误差的误差传递公式。

对于标准误差的传递则有:

$$\sigma_y = \sqrt{\left(\frac{\partial y}{\partial x_1}\right)^2 \sigma_{x_1}^2 + \left(\frac{\partial y}{\partial x_2}\right)^2 \sigma_{x_2}^2 + \cdots + \left(\frac{\partial y}{\partial x_n}\right)^2 \sigma_{x_n}^2} \tag{2-29}$$

式中,σ_{x_1}, σ_{x_2} 等为直接测量值的标准误差;σ_y 为间接测量值的标准误差。

计算函数的误差的各种关系式见表 2-1。

表 2-1 函数式的误差传递公式

函数式	误差传递公式	
	最大绝对误差 Δy	最大相对误差 $E_r(y)$
$y = x_1 + x_2 + \cdots + x_n$	$\Delta y = \pm(|\Delta x_1| + |\Delta x_2| + \cdots + |\Delta x_n|)$	$E_r(y) = \Delta y/y$
$y = x_1 - x_2$	$\Delta y = \pm(|\Delta x_1| + |\Delta x_2|)$	$E_r(y) = \Delta y/y$
$y = x_1 x_2$	$\Delta y = \pm(|x_1 \Delta x_2| + |x_2 \Delta x_1|)$	$E_r(y) = \pm\left(\left|\frac{\Delta x_1}{x_1}\right| + \left|\frac{\Delta x_2}{x_2}\right|\right)$
$y = x_1 x_2 x_3$	$\Delta y = \pm(|x_1 x_2 \Delta x_3| + |x_1 x_3 \Delta x_2| + |x_2 x_3 \Delta x_1|)$	$E_r(y) = \pm\left(\left|\frac{\Delta x_1}{x_1}\right| + \left|\frac{\Delta x_2}{x_2}\right| + \left|\frac{\Delta x_3}{x_3}\right|\right)$

续表

函数式	误差传递公式	
	最大绝对误差 Δy	最大相对误差 $E_r(y)$
$y = x^n$	$\Delta y = \pm (|nx^{n-1}\Delta x|)$	$E_r(y) = \pm \left(n\left|\dfrac{\Delta x}{x}\right|\right)$
$y = \sqrt[n]{x}$	$\Delta y = \pm \left(\left|\dfrac{1}{n}x^{\frac{1}{n}-1}\Delta x\right|\right)$	$E_r(y) = \pm \left(\dfrac{1}{n}\left|\dfrac{\Delta x}{x}\right|\right)$
$y = x_1/x_2$	$\Delta y = \pm \left(\dfrac{|x_2 \Delta x_1| + |x_1 \Delta x_2|}{x_2^2}\right)$	$E_r(y) = \pm \left(\left|\dfrac{\Delta x_1}{x_1}\right| + \left|\dfrac{\Delta x_2}{x_2}\right|\right)$
$y = cx$	$\Delta y = \pm |c \Delta x|$	$E_r(y) = \pm \left(\left|\dfrac{\Delta x}{x}\right|\right)$
$y = \lg x$	$\Delta y = \pm \left|0.4343 \dfrac{\Delta x}{x}\right|$	$E_r(y) = \Delta y / y$
$y = \ln x$	$\Delta y = \pm \left|\dfrac{\Delta x}{x}\right|$	$E_r(y) = \Delta y / y$

二、实验数据处理

实验数据处理就是以测量为手段，以研究的概念、状态为基础，以数学运算为工具，推断出某测量值的真值，并导出某些具有规律性结论的整个过程。因此对实验数据进行处理，可使人们清楚地观察到各变量之间的定量关系，以便进一步分析实验现象，得出规律，指导生产与设计。

1. 列表法

为了数据处理的方便、不遗漏、有条理，需要对实验数据做初步整理。要求根据实验内容预先设计好记录及计算的表格。

实验数据记录表一般可分为下列几栏：实验测定数据栏（原始数据栏）、中间计算栏及实验结果栏。具体内容视实验内容而定。

在拟定记录表格时应注意下列问题：

① 各标题栏目必须标明物理量的名称和单位，名称应尽量用符号表示，单位及其数量级写在该符号的标题栏中。

② 记录的位数，应限于有效数字，同一栏中的有效数字的位数相同。

③ 对于数量级很大或很小的数，在标题栏中乘以适当的倍数，在数据栏中就可记为整数。

例如：$Re = 25000 = 2.5 \times 10^4$

标题栏中记为 $Re \times 10^{-4}$，数据栏中可记为 2.5。

④ 由左至右，按实验测定数据、中间计算数据及实验结果等次序排列，这样便于记录及计算整理。当然，实验测定数据与中间计算数据、实验结果等栏分开也可以。

为使运算结果由繁到简，节省时间，减少差错，尽量采用常数归纳法，预先算出各常数项。

2. 图示法

上述列表法，一般难见数据的规律性。若将实验结果综合表中的自变量和因变量数据标绘在坐标系上，用图形来表示因变量和自变量的依从关系，简明直观，易于显示出结果的规律性或变化趋势，更便于比较。

在工程实验中正确作图必须遵循的基本原则如下：

① 对于两个变量的系统，习惯上选横轴为自变量，纵轴为因变量。在两轴侧要标明变量名称、符号和单位。尤其是单位，初学者往往受纯数字的影响容易忽略。

② 坐标的分度的选择，要反映出实验数据的有效数字位数，即与被标数值精度一致，并要求方便易读。坐标分度值不一定从零开始，而使图形占满全幅坐标系为合适。

坐标分度是指沿 x、y 轴每条坐标轴所代表数值的大小，实质上是选择坐标的比例尺。对于同一组数据，由于所选择的比例尺不同，往往从所绘得的图形中得出完全不同的结论。例如：确定 y 与 x 的函数关系时，得到表 2-2 所列实验数据。

表 2-2 实验数据

x	1.00	2.00	3.00	4.00
y	8.0	8.2	8.3	8.0

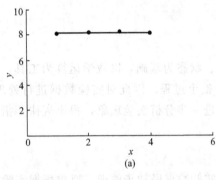

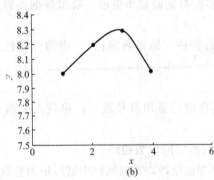

图 2-2 $y=f(x)$ 的函数关系

将这些数据绘制在图 2-2 中，图 2-2（a）与（b）的纵坐标比例不同，所表现的图形完全不同。从图 2-2（a）中的曲线可认为 y 与 x 无关，或者 y 为常数，而从图 2-2（b）则可以看出 y 是随 x 而变化的，由这两个图的结果看出，似乎所选择的坐标尺寸不同，函数的性质不同。事实上并非如此，由数学知识知，函数关系仅取决于自变量和因变量的值。函数的性质，是由函数的本质所决定，比例尺寸选择得不够恰当，只是不能揭示出其内部规律，并不能改变函数的本质。在图 2-2（a）与（b）中，图 2-2（a）的比例尺寸与所进行实验的精确度不相适应，因而不能描述出 y 与 x 的变化规律，因此在选定比例尺寸时，必须考虑实验的精确度。

若 y 及 x 的实验精度分别为 Δy 及 Δx，而点在矩形 $2\Delta y$，$2\Delta x$ 内波动，正确的作图法，要使

$$m_x \times 2\Delta x = m_y \times 2\Delta y \qquad (2\text{-}30)$$

式中，m_x，m_y 分别为 x 及 y 的尺系数，即每个刻度的长度（毫米/单位）。

在图 2-2 中，如果 y 及 x 的精度分别为 $\Delta y = \pm 0.2$ 及 $\Delta x = \pm 0.05$，则

$$\frac{\Delta y}{\Delta x} = \frac{0.2}{0.05} = 4.0 \qquad (2\text{-}31)$$

若取 $\qquad m_x = 4.0$（毫米/单位）

则 $\qquad m_y = 1.0$（毫米/单位）

③ 若在同一坐标系上，同时标绘几组测量值，则各组要用不同符号（如：•，×，△等），以示区别。若 n 组不同函数同绘在一坐标系上，则在曲线上要标明函数关系或名称，

或标明读数方向箭头。

④ 实验曲线以直线最易标绘，使用也最为方便。因此在处理数据时，尽量使曲线直线化。为此，根据不同情况将变量加以变换或选用不同坐标系，如在化学工程实验数据处理上，经常采用单对数和双对数坐标系。并且坐标是根据拟合直线的斜率接近于1做分度。

下列三种函数型式，可分别在不同坐标系上得到直线图形。

① 直线函数型　变量 x，y 间的函数关系式为

$$y = a + bx \tag{2-32}$$

② 指数函数型　若变量 x，y 间存在指数函数型关系，则有

$$y = k e^{mx} \tag{2-33}$$

式中，k，m 为待定系数。

在这种情况下，若把 x，y 数据在直角坐标系上作图，所得图形必为一曲线。若对式(2-33)两边取对数，则

$$\lg y = \lg k + mx \lg e \tag{2-34}$$

令

$$\lg y = Y$$
$$m \lg e = b_1$$
$$\lg k = a_1$$

则式 (2-34) 变为

$$Y = a_1 + b_1 x \tag{2-35}$$

经上述处理，x、Y 变成线性关系，以 $\lg y = Y$ 对 x 在直角坐标系上作图，其图形也是直线。

为了避免对每一个实验数据 Y 都取对数的麻烦，可采用单对数坐标系。该坐标系的一个轴为对数刻度，另一轴的刻度仍为直角坐标。把实验数据标识在这种坐标系上，若为直线者关联式必为指数函数型。

③ 幂函数型　若变量 x，y 间存在幂函数型关系式，则有

$$y = k x^m \tag{2-36}$$

式中，k，m 为待定系数。

式 (2-36) 直接在直角坐标系上作图必为曲线，为此，把式 (2-36) 两边取对数，则

$$\lg y = \lg k + m \lg x \tag{2-37}$$

令

$$\lg y = Y$$
$$\lg x = X$$
$$m = b_2$$
$$\lg k = a_2$$

则式(2-37) 变为

$$Y = a_2 + b_2 X \tag{2-38}$$

根据式 (2-38)，把实验数据 x、y 取对数 $\lg y = Y$、$\lg x = X$，在直角坐标系上作图也得一直线。当然也可直接采用双对数坐标系，所得结果完全相同。使用双对数坐标系，需注意：

a. 标在对数坐标系上的数是真数而不是对数。

b. 由于 $\lg 1 = 0$，与普通坐标系上 $x = 0$ 相当的一条直线，在对数坐标系上则为 $x = 1$ 的直线。

c. 确定指数 m 与系数 k 时注意，m 不能采用在普通坐标上的计算方法，而是

$$m = \frac{\text{两点在 } y \text{ 轴上的距离}}{\text{两点在 } x \text{ 轴上的距离}} \neq \frac{y \text{ 轴读数差}}{x \text{ 轴读数差}}$$

k 即为该直线与 $x=1$ 直线的交点的 y 值，也可在直线上任取一组 (x_1, y_1) 值代入式(2-36)而求得。也可采用平均值法或最小二乘法来确定系数 k、m 值。

对于非线性关系的实验点，应将其采用合理的方法拟合成光滑的曲线，对于离散的实验点大致取平均值作图，显而易见，这种方式误差较大，精确度低。

对于线性关系的实验点，可采用最小二乘法作图，这样求得的曲线，误差最小，精确度较高。

若变量多于两个时，如 $y = f(x, z)$ 在作图时，先固定一个变量（例如使 z 固定），求出 $y \sim x$ 关系，这样可得不同 z 值下的一组图线。

3. 实验数据数学方程表示法

当一组实验数据用列表法和图示法表示后，在某些场合常需进一步用数学方程来描述各个参数和变量之间的关系。用数学公式表示变量之间的相互关系，不但简单，而且使用也方便。其方法就是将实验中得到的数据绘制成曲线，与已知函数关系式的典型曲线（直线方程、指数方程、抛物线方程、圆及椭圆方程等）进行对照，加以选择，同时求出方程的常数和系数，这样，经验公式就可以求得。

经验公式中求常数和系数的方法很多，最常用的是直线图解法、平均值法和最小二乘法。

(1) 用直线图解法求待定系数

当所研究的函数关系是线性的，或者可以利用直线化方法化为线性时，均可用式 $y = a + bx$ 表达。该直线的斜率（$\Delta y / \Delta x$），即为方程中 b 值。直线在 y 轴上的截距 a，即为方程中的 a 值。

特别要注意在对数坐标系上直线方程的斜率和截距的求取与直角坐标系上的不同。

(2) 用平均值法求待定系数

选择能使各测定值的偏差的代数和为零的那条曲线为理想曲线。现假定得出的理想曲线为直线，设其方程为

$$y = a + bx \tag{2-39}$$

设测定值为 x_i、y_i，将 x_i 代入式(2-39)，所得的 y 值为 y_i'，则

$$y_i' = a + bx_i \tag{2-40}$$

理论上，$y_i' = y_i$。然而，一般由于测定误差，实测点偏离直线，因而 $y_i' \neq y_i$。若设 y_i 和 y_i' 的差为 Δi，则

$$\Delta i = y_i - y_i' = y_i - (a + bx_i) \tag{2-41}$$

最理想曲线是使这个差值的总和为零的直线。设测定值的个数为 N

$$\sum \Delta i = \sum y_i - Na - b \sum x_i = 0 \tag{2-42}$$

由式(2-42)定出 a、b，则以 a、b 为常数的直线即为所求的理想曲线。

由于式(2-42)含有两个未知数 a 和 b，所以需把测定值按实验数据的次序分成相等和近似相等的两组，分别建立相应的方程式，把这两个方程式联立，解方程组即得 a、b。

(3) 用最小二乘法求待定系数

偏差有"正"有"负"，数据处理时，正、负可能相抵消，而不足以表示数值偏差的实

质，但偏差的平方均为正值，若偏差的平方和最小，即各偏差最小。最小二乘法是这样定义的：最理想的曲线就是能使各点同曲线的偏差的平方和最小。它是根据误差理论得出的。根据式（2-41）可知

$$\Delta i^2 = (y_i - y_i')^2 = [y_i - (a + bx_i)]^2 \tag{2-43}$$

取得最佳值的条件是

$$\sum \Delta i^2 = \sum [y_i - (a + bx_i)]^2 \to 最小 \tag{2-44}$$

式（2-44）对 a 和 b 的偏微分值同时为 0 时，这个条件就可得到满足。

$$\frac{\partial(\sum \Delta i^2)}{\partial a} = -2\sum [y_i - (a + bx_i)] = 0$$

所以

$$\sum y_i = Na + b\sum x_i \tag{2-45}$$

$$\frac{\partial(\sum \Delta i^2)}{\partial b} = -2\sum x_i [y_i - (a + bx_i)] = 0$$

所以

$$\sum x_i y_i = a\sum x_i + b\sum x_i^2 \tag{2-46}$$

式（2-45）和式（2-46）是用于最小二乘法求直线式中常数 a 和 b 时的一般公式。
用行列式求解：

$$b = \frac{\begin{vmatrix} \sum y_i & N \\ \sum x_i y_i & \sum x_i \end{vmatrix}}{\begin{vmatrix} \sum x_i & N \\ \sum x_i^2 & \sum x_i \end{vmatrix}} = \frac{\sum x_i \sum y_i - N\sum x_i y_i}{(\sum x_i)^2 - N\sum x_i^2} \tag{2-47}$$

$$a = \frac{\begin{vmatrix} \sum x_i & \sum y_i \\ \sum x_i^2 & \sum x_i y_i \end{vmatrix}}{\begin{vmatrix} \sum x_i & N \\ \sum x_i^2 & \sum x_i \end{vmatrix}} = \frac{\sum x_i \sum x_i y_i - \sum y_i \sum x_i^2}{(\sum x_i)^2 - N\sum x_i^2} \tag{2-48}$$

由于最小二乘法的偏差平方和最小，其偶然误差总是最小的。但此法手算比较繁琐，若利用计算机则无此缺点。当实验数据能很好地符合一条直线时，平均值法与图解法均可使用。

求数学方程中常数除上面介绍的直线图解法、平均值法和最小二乘法外，还可用差分法、选点法及回归分析法等。

4. 相关系数 r 及显著性检验

（1）相关系数 r

实验数据的变量之间关系具有不确定性，一个变量的每一个值对应的是整个集合值。当 x 改变时，y 的分布也以一定的方式改变。在这种情况下，变量 x 和 y 间的关系就称为相关关系。

在求回归方程的过程中，并不需要先假定两个变量之间一定具有相关关系。即使平面上一堆完全杂乱无章的散点，也可用最小二乘法给定一条直线来表示 x 与 y 之间的关系。然而，这是毫无意义的。只有两个变量是线性关系才适宜线性回归。因此，必须有一个数量性

指标来描述两个变量线性关系的密切程度。对此引进一个叫相关系数 r 的统计量,用来判断两个变量之间的线性相关程度。

$$r = \frac{\sum (x_i - \bar{x})(y_i - \bar{y})}{\sqrt{\sum (x_i - \bar{x})^2 \sum (y_i - \bar{y})^2}} \tag{2-49}$$

式中

$$\bar{x} = \frac{\sum x_i}{n}$$

$$\bar{y} = \frac{\sum y_i}{n}$$

在概率中可以证明,任意两个随机变量的相关关系的绝对值不大于1,即:

$$|r| \leqslant 1 \text{ 或 } 0 \leqslant |r| \leqslant 1 \tag{2-50}$$

r 的物理意义:它表明两个随机变量 x 与 y 的线性相关程度。现分几种情况加以说明:

① 当 $r = \pm 1$ 时,即 n 组实验值 (x_i, y_i) 全部落在直线 $y = a + bx$ 上,此时称完全相关。

② 当 $0 < |r| < 1$ 时,代表绝大多数的情况,这时 x 与 y 存在着一定线性关系。当 $r > 0$、$b > 0$ 时,散点图的分布是 y 随 x 增加而增加,此时称 x 与 y 正相关。当 $r < 0$、$b < 0$ 时,y 随 x 增加而减少,称 x 与 y 负相关。$|r|$ 越小,离散点距回归线越远,越分散。当 $|r|$ 越接近1时,即 n 组实验值 (x_i, y_i) 越靠近 $y = a + bx$,变量 y 与 x 之间的关系,越接近于线性关系。

③ 当 $r = 0$,变量之间就完全没有线性关系了,应该指出,没有线性关系,不等于不存在其他函数关系。

(2) 显著性检验

如上所述,相关系数 r 的绝对值越接近1,x、y 间越线性相关。但究竟 $|r|$ 接近到什么程度才能说明 x 与 y 之间存在线性相关关系呢?这就有必要对相关系数进行显著性检验。只有当 $|r|$ 达到一定程度才可用回归直线来近似地表示 x、y 之间的关系。此时,可以说相关关系显著。一般来说,使相关显著的 r 值与实验数据点的个数 n 有关。只有 $|r| > r_{\min}$ 时,才能采用线性回归方程来描述其变量之间的关系。$r_{\min}$ 值可见相关系数显著性检验表(表2-3)。利用该表可根据实验数据点个数 n 及显著水平 a 查出相应的 $r_{\min}$。一般可取 $a = 1\%$ 或 $a = 5\%$。

当实验数据点个数为9时,若实际的 $|r| \geqslant 0.798$,则说明该线性相关关系在 $a = 0.01$ 水平上显著。若 $0.798 \geqslant |r| \geqslant 0.666$,则说该线性相关关系在 $a = 0.05$ 水平上显著。若实际的 $|r| \leqslant 0.666$,则说明相关关系不显著,此时认为 x、y 线性不相关,配回归直线毫无意义。a 越小则显著程度越高。

表2-3 相关系数显著性检验表

自由度 ($n-2$)	显著性水平		自由度 ($n-2$)	显著性水平		自由度 ($n-2$)	显著性水平	
	5%	1%		5%	1%		5%	1%
1	0.997	1.000	5	0.754	0.874	9	0.602	0.735
2	0.950	0.990	6	0.707	0.834	10	0.576	0.708
3	0.878	0.959	7	0.666	0.798	11	0.553	0.684
4	0.811	0.917	8	0.632	0.765	12	0.532	0.661

续表

自由度 ($n-2$)	显著性水平		自由度 ($n-2$)	显著性水平		自由度 ($n-2$)	显著性水平	
	5%	1%		5%	1%		5%	1%
13	0.514	0.641	24	0.388	0.496	60	0.250	0.325
14	0.497	0.623	25	0.381	0.487	70	0.232	0.302
15	0.482	0.606	26	0.374	0.478	80	0.217	0.283
16	0.468	0.590	27	0.367	0.470	90	0.205	0.267
17	0.456	0.575	28	0.361	0.463	100	0.195	0.254
18	0.444	0.561	29	0.355	0.456	125	0.174	0.228
19	0.433	0.549	30	0.349	0.449	150	0.159	0.208
20	0.423	0.537	35	0.325	0.418	200	0.138	0.181
21	0.413	0.526	40	0.304	0.393	300	0.113	0.148
22	0.404	0.515	45	0.288	0.372	400	0.098	0.128
23	0.396	0.505	50	0.273	0.354	1000	0.062	0.081

第三章 化工原理虚拟仿真实验概述

一、软件运行环境

操作系统：Windows XP/7/8
CPU：Intel 双核 @ 2.40GHz 或以上
内存：2G 以上
显卡：显存 1G 以上，位宽 256bit 以上

二、化工原理实验虚拟仿真系统启动

在一台正常运行的计算机上，双击安装包，即可安装虚拟仿真软件。

以填料塔吸收过程实验为例：

吸收实验仿真系统的开始界面如图 3-1 所示，其中可在 Screen resolution 中选择所需要的屏幕分辨率，在 Graphics quality 中选择所需的图像质量。

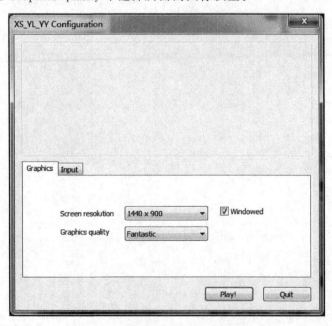

图 3-1 吸收实验仿真系统开始界面

三、化工原理实验虚拟仿真菜单功能

当鼠标光标移动到屏幕上侧区域时,会自动触发隐匿式菜单,如图3-2所示。

图 3-2　PC 端隐匿式菜单

1. 任务

任务菜单中包括了新手上路、注意事项、实验概述、结构认知、案例教学等功能,单击该选项会跳出如图3-3所示的层级菜单。

图 3-3　PC 端层级菜单

① 新手上路　对于未使用过虚拟仿真软件的同学,使用新手上路来了解软件的基本操作。

② 注意事项　选择该功能时会跳出相应的介绍界面,如图3-4所示。跳出该界面后,也会启动相应的语音功能,同时进行语音介绍。

③ 实验概述　选择该功能时会跳出相应的介绍界面,如图3-5所示。跳出该界面后,也会启动相应的语音功能,同时进行语音介绍。

④ 结构认知　选择流体力学综合实验仪器介绍后,可对流体力学综合实验有一个总体的结构认知,如图3-6所示为结构认知效果示意图。

⑤ 案例教学　点击案例教学任务选择界面,如图3-7所示。

选择相应案例以后,首先是对实验一些简要说明介绍,如图3-8所示。

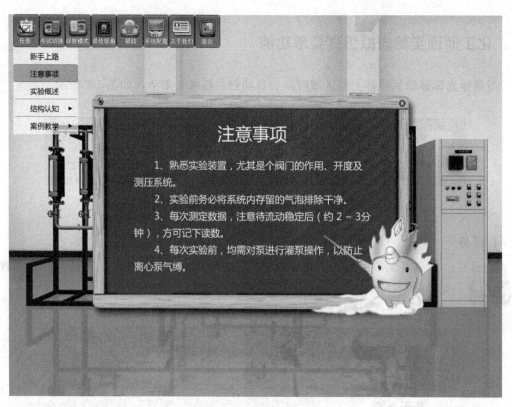

图 3-4　注意事项界面

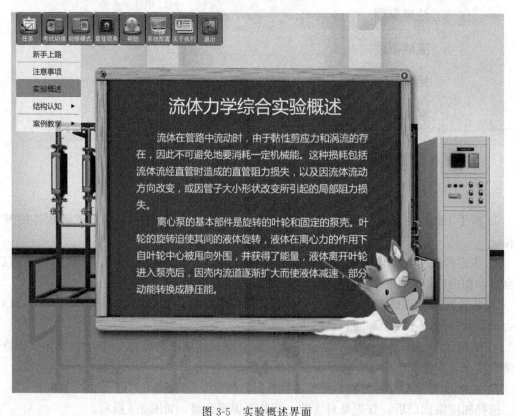

图 3-5　实验概述界面

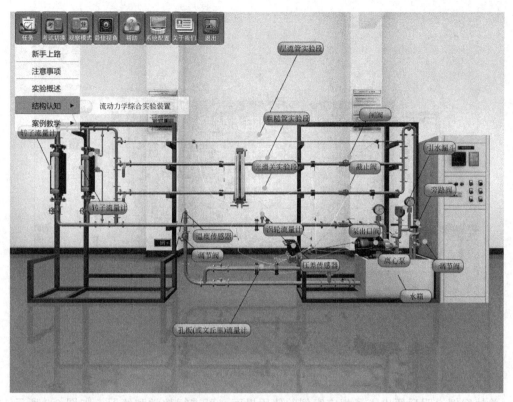

图 3-6　结构认知示意图

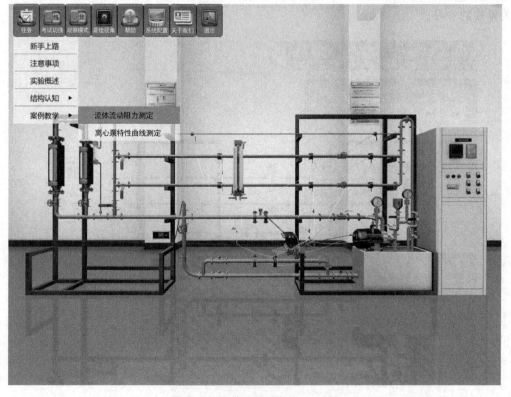

图 3-7　案例教学任务选择界面

第三章　化工原理虚拟仿真实验概述

图 3-8　案例说明界面

关掉案例说明后便进入了相应案例的使用界面,可进行教学和练习,如图 3-9 所示。根据所处功能阶段的不同,会跳出不同的功能菜单。用户可以根据相应文字和语音提示,进行相关操作的学习。

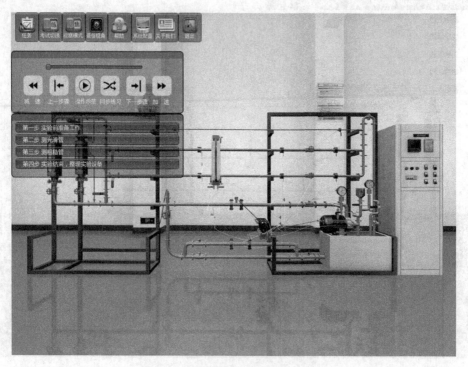

图 3-9　案例教学和练习功能界面

在教学过程进行时，所有交互功能都暂时屏蔽，程序会自动进行案例的相关操作，并伴有操作文字提示和语音提示，用户可以据此学习案例的操作流程。

在练习过程进行时，软件会根据当前步骤自动开放相应的功能，指导用户一步步学习案例的操作流程。

2. 考试切换

考试切换

当用户熟悉案例的操作流程后，可选择隐匿式菜单中的练考切换键，进入考试模式进行相应案例的考试过程。当打开程序后第一次切换到考试功能时，系统会弹出考试信息输入窗口要求用户输入姓名和学号，如图 3-10 所示。

在输入完信息后，则进入案例考试界面，如图 3-11 所示。此时所有功能开放，用户可进行自由操作，系统会实时监视用户的当前状态，根据用户的当前操作对其进行打分。

当用户无法完成当前步骤时，可选择"下一阶段"，自动放弃当前得分点，进入下一步骤的考核。当选择"交卷"时，系统会跳出确认菜单，确认后软件会自动将考试成绩发送到系统桌面，用户可以自行查看。

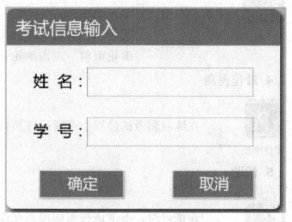

图 3-10 考试信息输入窗口

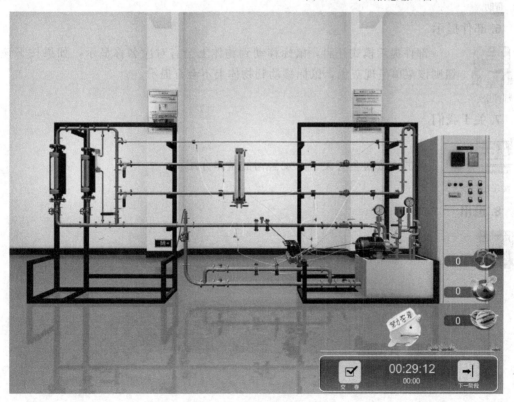

图 3-11 案例考试界面

3. 观察模式

该功能只在练和考情况下有效（有些进场景没有动作则按一下该键），模式分为"焦点模式"和"场景模式"，可以在此模式下"漫游"该实验室。

场景模式

场景模式下，采用键盘转换镜头：W，A，S，D 进行上下左右移动，Q，E 进行左旋和右旋，Z，X 进行视角上下调整，空格键用于跳跃。

场景模式	前	后	左	右	上	下
	W	S	A	D	Z	X

焦点模式

焦点模式下，采用鼠标转换镜头：按住鼠标中间可以旋转场景，同时按住中键和右键可平移场景。

焦点模式	前后	镜头方向	位置移动
	滚轮滚动	点击滚轮，移动鼠标	点击滚轮和右键，移动鼠标

4. 最佳视角

最佳视角

在练习和考试过程中可通过该按钮一键回到最佳位置。

5. 帮助

帮助

在练习时，点击该按钮即可高亮即将要操作的物体。

6. 部件提示

部件提示

部件提示模式开启，鼠标移动到物体上会有对应名称显示，如果按下该按钮则该菜单呈现灰色，鼠标移动到物体上不会有提示。

7. 关于我们

关于我们

点击该按钮会有关于开发公司的基本介绍。

8. 退出

退出

选择退出时，会弹出退出。

第四章

流体力学综合实验——管内流动阻力测定实验

一、实验目的和要求

1. 掌握测定流体流经直管、管件（阀门）时阻力损失的一般实验方法。
2. 测定直管摩擦系数 λ 与雷诺数 Re 的关系，验证在一般湍流区内 λ 与 Re 的关系曲线。
3. 测定流体流经管件（阀门）时的局部阻力系数 ξ。
4. 认识组成管路的各种管件、阀门，并了解其作用。

二、实验原理

不可压缩流体在管路中作稳定流动时，由于黏性剪应力和涡流的存在，消耗一定机械能。这种机械能损耗包括流体流经直管时造成的直管阻力损失，以及因流体流经管件，如阀门、弯头、变径管、三通等，流道的截面和流体流动方向改变所引起的局部阻力损失。流体流动的总阻力为直管阻力与局部阻力之和。

1. 流量计校核

可以通过体积计时或重量计时对流量计读数进行校核。

2. 雷诺数 Re

$$Re = \frac{du\rho}{\mu} \tag{4-1}$$

$$u = \frac{V}{900\pi d^2} \tag{4-2}$$

式中　Re——雷诺数，无量纲；
　　　d——直管内径，m；
　　　u——流体在管内流动的平均流速，m/s；
　　　ρ——流体密度，kg/m³；
　　　μ——流体黏度，kg/(m·s)；
　　　V——采用涡轮流量计测流体流量，m³/h。

3. 直管阻力摩擦系数 λ 的测定

流体在水平等径直管中稳定流动时，直管阻力与流体流动（雷诺数）及管路（管内壁粗糙度）损失为：

$$h_f = \frac{\Delta p_f}{\rho} = \frac{p_1 - p_2}{\rho} = \lambda \frac{l}{d} \times \frac{u^2}{2} \tag{4-3}$$

即

$$\lambda = \frac{2d\Delta p_f}{\rho l u^2} \tag{4-4}$$

式中　λ——直管阻力摩擦系数，无量纲；

Δp_f——流体流经 l(m) 直管的压力降，Pa；

h_f——单位质量流体流经 l(m) 直管的机械能损失，J/kg；

l——直管长度，m。

4. 局部阻力系数 ξ 的测定

如采用阻力系数法：流体通过某一管件（如阀门）时的机械能损失可表示为流体在小管径内流动时平均动能的某一倍数。即

$$h_f' = \frac{\Delta p_f'}{\rho} = \xi \frac{u^2}{2} \tag{4-5}$$

故

$$\xi = \frac{2\Delta p_f'}{\rho u^2} \tag{4-6}$$

式中　h_f'——单位质量流体流经某一管件或阀门时的机械能损失，J/kg；

ξ——局部阻力系数，无量纲；

$\Delta p_f'$——局部阻力的压力降，Pa。

局部阻力压力降的测量方法：因管件对流经管件前后的流体流动均有影响，故需要测量包含管件和直管（总长度 l'）在内总的压降 $\sum \Delta p$，再减去直管部分的压降，方能获得实际管件所产生的局部阻力，而直管段的压降可通过直管（长度 l）阻力（Δp_f）实验结果求取。

$$\Delta p_f' = \sum \Delta p - \frac{l'}{l} \Delta p_f \tag{4-7}$$

式中　$\sum \Delta p$——包含管件（阀门）与直管（长度为 l'）的压降，Pa；

l'——$\sum \Delta p$ 测压点（管件两侧）直管的长度，m。

三、实验装置和流程

1. 实验装置

实验装置（见图 4-1）由水箱，离心泵，不同管径、材质的管道，阀门，待测管件，流量计和压差传感器等组成。管路有三段并联长直管，自上而下分别用于测定直管层流阻力、粗糙管直管阻力系数、光滑管直管阻力系数。同时在粗糙管直管和光滑管直管上分别装有闸阀和截止阀，以此作为待测管件，用于测定不同种类阀门的局部阻力系数。虚拟仿真示意图

见图 4-2。

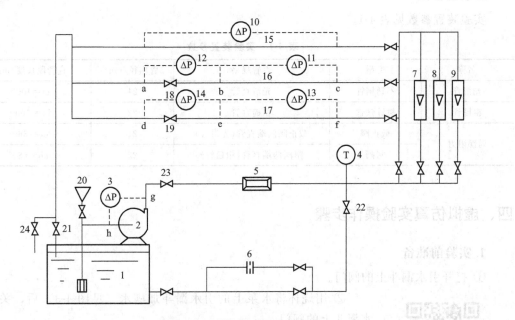

图 4-1 流体力学综合实验装置

1—水箱；2—离心泵；3，10~14—压差传感器；4—温度计；5—涡轮流量计；6—孔板（或文丘里）流量计；
7~9—转子流量计；15—层流管实验段；16—粗糙管实验段；17—光滑管实验段；18—闸阀；19—截止阀；
20—引水漏斗；21，22—流量调节阀；23—泵出口阀；24—旁路阀；a~h—取压点

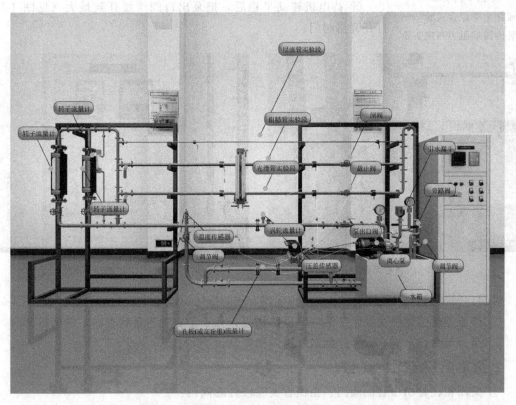

图 4-2 流体力学综合实验装置虚拟仿真示意图

第四章 流体力学综合实验——管内流动阻力测定实验

2. 装置参数

实验装置参数见表 4-1。

表 4-1 实验装置参数

名称	类型	直管规格	管内径/mm	直管段长度/mm
光滑管	不锈钢管	光滑直管	21	ef=1000
粗糙管	镀锌铁管	粗糙直管	22	bc=1000
局部阻力	截止阀	截止阀两端直管（光滑管）	21	de=660
	闸阀	闸阀两端直管（粗糙管）	22	ab=680

四、虚拟仿真实验操作步骤

1. 实验前准备

① 打开引水漏斗上的阀门。

扫一扫动画
流体力学综合实验——
管内流动阻力测定实验

② 用烧杯将水泵上的引水漏斗加满水（见图 4-3）后，关闭引水漏斗上的阀门。

③ 关闭泵出口阀。

④ 开启仪表柜上的总电源开关（见图 4-4），仪表电源开关，启动水泵电机。

⑤ 待电机转动平稳后，把泵出口阀缓缓开到最大（见图 4-5），打开管路阀门。

图 4-3 离心泵灌水

图 4-4 控制柜操作

⑥ 打开管路上的引压管阀门及压差传感器后侧的排气阀，对引压管进行排气（见图 4-6）。

⑦ 待压差传感器引压管路中的气泡全部排完，处于稳定的测量状态时，关闭排气阀。

⑧ 全开流量调节阀，以排除测试管路内的空气。

2. 测光滑管及管件（截止阀）

① 关闭粗糙管引压管的阀门、粗糙管实验段处总阀门。

② 打开光滑管实验段处总阀门、光滑管引压管的阀门，测量光滑管实验段。

化工原理实验及虚拟仿真（双语）

图 4-5　打开泵出口阀

图 4-6　打开压差传感器后侧的排气阀

③ 确认该管路上的管件（截止阀）完全打开，全开流量调节阀。
④ 待显示屏上流量数据稳定时，读出此时管路的最大流量。
⑤ 根据管路的最大流量分配流量，调节流量调节阀（见图 4-7）。
⑥ 流量改变后，待流动达到稳定后，记录数据（见图 4-8）。

图 4-7　开启流量调节阀

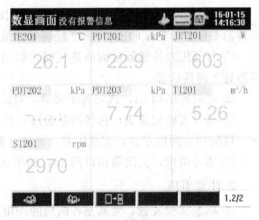

图 4-8　数据显示

3. 测粗糙管及管件（闸阀）

① 关闭光滑管引压管的阀门、光滑管实验段处总阀门。
② 打开粗糙管实验段处总阀门（见图 4-9）、粗糙管引压管的阀门，测量粗糙管实验段。
③ 确认该管路上的管件（闸阀）完全打开（见图 4-10），全开流量调节阀。
④ 待显示屏上流量数据稳定时，读出此时管路的最大流量。
⑤ 根据管路的最大流量分配流量，调节流量调节阀（见图 4-7）。
⑥ 流量改变后，待流动达到稳定后，记录数据（见图 4-8）。

4. 实验结束，整理实验设备

① 关闭流量调节阀。
② 关闭引压管的阀门、实验段处总阀门。
③ 关闭管路总阀门、泵出口阀。
④ 关闭水泵电机、仪表电源开关、总电源开关。

图4-9 打开粗糙管阀门

图4-10 确认管路上的闸阀完全打开

五、实验方法及步骤

1. 实验操作步骤

① 通过引水漏斗20给离心泵灌水,关闭泵出口阀23,打开电源,启动水泵电机,待电机转动平稳后,把泵的出口阀23缓缓开到最大。

② 打开压差传感器后面的排气阀,对引压管进行排气,完成后关闭排气阀,使压差传感器处于测量状态。

③ 开启旁路阀24,选定若干流量,对流量计做流量校核实验。

④ 设计实验记录表,选择测量管路,打开流量调节阀21,合理分配流量,每次改变流量,待流动达到稳定后,记录压差、流量、温度等数据。

⑤ 实验结束,关闭泵出口阀23,关闭水泵电机,切断水泵电源。

2. 注意事项

① 熟悉实验装置,尤其是各阀门的作用及测压系统的连接点。

② 实验记录数据前务必将引压管内存留的气泡排除干净。

③ 每次测定数据,注意需待流动稳定后,方可记下读数。

六、实验报告

1. 通过实验数据计算对应流量下的雷诺数 Re、光滑管和粗糙管的直管阻力系数 λ 以及管件(阀门)的局部阻力系数 ξ。

2. 根据光滑管、粗糙管的实验结果,在双对数坐标系上分别拟合出 $\lambda \sim Re$ 的关系曲线,对照相关手册上的 Moody 图,估算实验管路的相对粗糙度和绝对粗糙度。

3. 根据局部阻力的实验结果,分析局部阻力(阀门)的阻力系数 ξ 随雷诺数 Re 的变化情况,并与相关手册上推荐的经验值做比较。

4. 对以上实验结果的合理性进行分析讨论。

七、思考题

1. 如何检测管路中的空气已经被排除干净？
2. 以水做介质所测得的 $\lambda \sim Re$ 关系能否适用于其他流体？如何应用？
3. 在不同管路上、不同水温下测定的 $\lambda \sim Re$ 数据能否关联在同一条曲线上？
4. 如果测压口、孔边缘有毛刺或安装不垂直，对静压的测量有何影响？
5. 本实验装置上哪些压差的测量可以共用差压计？为什么？如何实现共用？

第五章

流体力学综合实验
——离心泵特性曲线测定实验

一、实验目的和要求

1. 了解离心泵结构与特性,掌握离心泵开、停车的正确操作方法和注意事项。
2. 测定离心泵在恒定转速下的操作特性,做出特性曲线。
3. 学习差压变送器、涡轮流量计、功率表等仪器仪表的工作原理和使用方法。

二、实验原理

在转速 n 固定不变的情况下,离心泵的实际扬程 H、功率消耗 N 及泵的效率 η 与泵送液能力 Q(即流量)之间的关系以曲线表示,称为离心泵的特性曲线,它能反映出泵的操作性能,可作为选择离心泵的依据。

离心泵的特性曲线可用下列三个函数关系表示:

$$H = f_1(Q); \quad N = f_2(Q); \quad \eta = f_3(Q)$$

这些函数关系均可由实验测得,其测定方法如下:

1. 流量 Q

流量 Q 通过流量计测量。

2. 泵的扬程 H

取离心泵进出口处测压点作为1、2两截面,机械能衡算式为:

$$z_1 + \frac{p_1}{\rho g} + \frac{u_1^2}{2g} + H = z_2 + \frac{p_2}{\rho g} + \frac{u_2^2}{2g} \tag{5-1}$$

因本实验泵进出口管径相同,故水流速相差不大,则速度平方差可忽略,则有

$$H = (z_2 - z_1) + \frac{p_2 - p_1}{\rho g} = H_0 + \frac{\Delta p}{\rho g} \tag{5-2}$$

$$H_0 = (z_2 - z_1)$$

$$\Delta p = (p_2 - p_1)$$

式中 H_0——泵进出口测压点的位差,m,本实验为 0.1m;

  ρ——流体(水)密度,kg/m³;

g——重力加速度，m/s²；

p_1，p_2——泵进口和泵出口的表压，Pa；

u_1，u_2——泵进、出口的流速，m/s；

z_1，z_2——真空表、压力表的安装高度，m；

Δp——泵的出口和进口之间的压差，Pa。

3. 泵的轴功率 N

轴功率 N 是单位时间内电机向离心泵轴输入的功。

$$N = N_电 k_电 k_传 \tag{5-3}$$

式中 $N_电$——泵的电机功率，由功率表测量，W；

$k_电$——电机效率，实验室提供；

$k_传$——传动效率，因电机与泵是联轴节传动，故 $k_传 = 1$。

4. 泵的效率 η

由于水力损失、容积损失和机械损失等造成泵内功率损失，有效功率 N_e 是单位时间内流体经过泵时所获得的实际功，泵的效率 η 是泵的有效功率 N_e 与轴功率 N 的比值。

泵的有效功率 N_e

$$N_e = HQ\rho g \tag{5-4}$$

泵的效率 η

$$\eta = \frac{N_e}{N} \times 100\% \tag{5-5}$$

5. 转速校核

泵的特性曲线是在一定转速下的实验测定所得。感应电动机在转矩改变时，随着泵流量 Q 的变化，泵的转速 n 也有所变化，因此在拟合泵的特性曲线之前，须将实测数据校核为某一定转速 n' 下（可取离心泵的额定转速）的数据。

流量

$$Q' = Q \frac{n'}{n} \tag{5-6}$$

扬程

$$H' = H \left(\frac{n'}{n}\right)^2 \tag{5-7}$$

轴功率

$$N' = N \left(\frac{n'}{n}\right)^3 \tag{5-8}$$

效率

$$\eta' = \frac{Q'H'\rho g}{N'} = \frac{QH\rho g}{N} = \eta \tag{5-9}$$

三、实验装置和流程

实验对象部分由贮水箱、离心泵、涡轮流量计和压差传感器等组成。

流体流量使用涡轮流量计进行测量，泵进出口压差采用压差传感器进行测量，泵的轴功率由功率表测量，流体温度采用Pt100温度传感器进行测量。

实验装置如图4-1、图4-2所示。

四、虚拟仿真实验操作步骤

1. 实验前准备工作

① 先打开引水漏斗上的阀门。

扫一扫动画
流体力学综合实验——
离心泵特性曲线测定实验

② 用烧杯将水泵上的引水漏斗加满水（见图4-3）后，关闭引水漏斗上的阀门。

③ 关闭泵出口阀。

④ 开启仪表柜上的总电源开关（见图4-4）、仪表电源开关。

⑤ 启动水泵电机。

⑥ 待电机转动平稳后，把泵出口阀缓缓开到最大（见图4-5）。

⑦ 打开压差传感器上引压管的阀门（见图5-1）。

⑧ 打开压差传感器后侧的排气阀，对引压管进行排气。

⑨ 待引压管中的气泡全部排完，处于稳定的测量状态时，关闭压差传感器后侧的排气阀。

⑩ 全开管路上的流量调节阀门，以排出测试管路内的空气（见图5-2）。

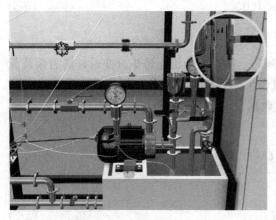

图5-1 打开引压管阀门

图5-2 开启流量调节阀

2. 实验测试开始

① 待显示屏上流量数据稳定时，读出此时管路的最大流量。

② 并记录最大流量时的压差、电机功率和流体温度。

③ 根据管路的最大流量分配流量，调节流量调节阀。

④ 流量改变后，待流动达到稳定后，读数并记录不同流量下的各数据。

⑤ 记录设备的相关数据，如离心泵型号、额定流量等。

3. 实验结束，整理实验设备

① 关闭流量调节阀。

② 关闭引压管的阀门、实验段处总阀门。

③ 关闭管路总阀门、泵出口阀。
④ 关闭水泵电机、仪表电源开关、总电源开关。

五、实验方法及步骤

1. 实验操作步骤

① 通过引水漏斗 20 给离心泵灌水，关闭泵出口阀 23，打开电源，启动水泵电机，待电机转动平稳后，把泵的出口阀 23 缓缓开到最大。
② 对压差传感器引压管进行排气，完成后关闭排气阀，使压差传感器处于测量状态。
③ 缓缓开启流量调节阀 22，合理分配流量，每次改变流量，待流动达到稳定后，记录压差、流量、电机功率、流体温度等数据。
④ 实验结束，关闭泵出口阀 23，关闭水泵电机，关闭仪表电源和总电源开关，将装置恢复原状。

2. 注意事项

① 每次实验前，均需对泵进行灌泵，以防止离心泵发生气缚。
② 启动离心泵时，注意关闭泵出口阀门，保护电机。
③ 离心泵启动后，注意防止泵轴高速旋转引起的人身伤害事故。

六、实验报告

1. 在同一坐标系拟合出一定转速下泵的特性曲线（$H \sim Q$，$N \sim Q$，$\eta \sim Q$）。
2. 分析实验结果，判断泵的最适宜的工作范围。

七、思考题

1. 从所测实验数据分析，离心泵在启动时为什么要关闭出口阀门？
2. 离心泵在启动前为何要引水灌泵？如果已经引水灌泵了，离心泵还是不能正常运行可能是什么原因？
3. 由实验记录数据知泵输送的水量越大，泵进口处的真空度也越大，为什么？
4. 泵启动后，出口阀如果不开，压力表读数是否会逐渐上升？为什么？
5. 正常工作的离心泵，能否在离心泵的进口管处安装调节阀？为什么？

第六章

对流传热系数的测定实验

一、实验目的和要求

1. 掌握空气在传热管内对流传热系数的测定方法，了解影响传热系数的因素和强化传热的途径。

2. 将实验数据整理成 $Nu=ARe^n$ 形式的特征数方程，并与手册中经验式进行比较。

3. 了解温度、加热功率、空气流量的自动控制原理和方法。

二、实验原理

在工业生产过程中，大量情况下，采用间壁式换热方式进行换热。所谓间壁式换热，就是冷、热两种流体之间有一固体壁面，两流体分别在固体壁面的两侧流动，不直接接触，通过固体壁面（传热元件）进行热量交换。

本装置主要研究汽-气换热，包括普通管和加强管。水蒸气和空气通过紫铜管间接换热，空气流经紫铜管内，水蒸气流经紫铜管外环隙，采用逆流换热。所谓加强管，是在紫铜管内加了弹簧，增大了绝对粗糙度，进而增大了空气的湍流程度，使换热效果更明显。

1. 间壁式传热基本原理

如图 6-1 所示，间壁式传热过程由热流体对固体壁面的对流传热、固体壁面的热传导和固体壁面对冷流体的对流传热所组成。

间壁式传热在传热过程达到稳态后，有

$$Q = m_1 c_{p1}(T_1 - T_2) = m_2 c_{p2}(t_2 - t_1)$$
$$= \alpha_1 A_1 (T - T_W)_m = \alpha_2 A_2 (t_W - t)_m$$
$$= KA\Delta t_m \tag{6-1}$$

热流体与固体壁面的对数平均温差可由式（6-2）计算：

$$(T - T_W)_m = \frac{(T_1 - T_{W1}) - (T_2 - T_{W2})}{\ln \dfrac{T_1 - T_{W1}}{T_2 - T_{W2}}} \tag{6-2}$$

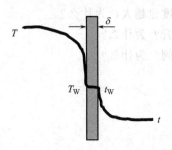

图 6-1 间壁式传热过程示意图

固体壁面与冷流体的对数平均温差可由式（6-3）计算：

$$(t_W - t)_m = \frac{(t_{W1} - t_1) - (t_{W2} - t_2)}{\ln \dfrac{t_{W1} - t_1}{t_{W2} - t_2}} \tag{6-3}$$

热、冷流体间的对数平均温差可由式（6-4）计算：

$$\Delta t_m = \frac{(T_1 - t_2) - (T_2 - t_1)}{\ln \dfrac{T_1 - t_2}{T_2 - t_1}} \tag{6-4}$$

式中　　　　　Q——传热量，W；

m_1，m_2——热流体、冷流体的质量流量，kg/s；

c_{p1}，c_{p2}——定性温度下热流体、冷流体的比热容，J/(kg·℃)；

T_1，T_2——热流体的进、出口温度，℃；

t_1，t_2——冷流体的进、出口温度，℃；

T_{W1}，T_{W2}——热流体进、出口端热流体侧的壁面温度，℃；

t_{W1}，t_{W2}——冷流体进、出口端冷流体侧的壁面温度，℃；

α_1，α_2——热流体、冷流体与固体壁面的对流传热系数，W/(m²·℃)；

A_1，A_2——热流体、冷流体侧的传热面积，m²；

$(T-T_W)_m$，$(t-t_W)_m$——热流体、冷流体与固体壁面的对数平均温差，℃；

K——以传热面积 A 为基准的总传热系数，W/(m²·℃)；

A——平均传热面积，m²；

Δt_m——冷、热流体的对数平均温差，℃。

本实验的热流体是蒸汽，冷流体是空气。

2. 空气流量的测定

空气在无纸记录仪上显示的体积流量，与空气流过孔板时的密度有关。考虑到实际过程中，空气的进口温度不是定值，为了处理上的方便，无纸记录仪上显示的体积流量是将孔板处的空气密度 ρ_0 当作 1kg/m³ 时的读数，因此，如果空气实际密度不等于该值，则空气的实际体积流量应按式（6-5）进行校正：

$$V' = \frac{V}{\sqrt{\rho_0}} \tag{6-5}$$

空气质量流量 m 可由式（6-6）计算：

$$m = V' \rho_0 \tag{6-6}$$

式中　V'——空气实际体积流量，m³/s；

V——无纸记录仪上显示的空气的体积流量，m³/s；

ρ_0——空气在孔板处的密度，kg/m³。

本实验中 ρ_0 即为空气在进口温度 t_1 下对应的密度。由于实验系统内压力变化不大，压力按照标准大气压计。

3. 空气在传热管内对流传热系数 α 的测定

(1) 牛顿冷却定律法

在本装置的套管加热器中，环隙内通水蒸气，紫铜管内通空气，水蒸气在紫铜管表面冷凝放热而加热空气。在传热过程达到稳定后，空气侧传热由式（6-1）可得：

$$Q = m_2 c_{p2}(t_2 - t_1) = \alpha_2 A_2 (t_W - t)_m \tag{6-7}$$

即

$$\alpha_2 = \frac{m_2 c_{p2}(t_2 - t_1)}{A_2 (t_W - t)_m} \tag{6-8}$$

t_{W1} 和 t_{W2} 分别是换热管空气进口处的内壁温度和空气出口处的内壁温度，当内管材料导热性能很好，即 λ 值很大，且管壁厚度较小时，可认为 $T_{W1} \approx t_{W1}$ 及 $T_{W2} \approx t_{W2}$，T_{W1} 和 T_{W2} 分别是空气进口处的换热管外壁温度（T3 和 T6）和空气出口处的换热管外壁温度（T2 和 T5），见流程图 6-3（横管）和图 6-4（竖管）。

一般情况下，直接测量固体壁面温度，尤其是管内壁温度，实验技术难度较大，因此，工程上也常通过测量相对较易测定的流体温度来间接推算流体与固体壁面间的对流传热系数。

下面介绍其他两种测定对流传热系数 α 的实验方法（2）和（3）。

(2) 近似法

以换热管内壁面积为基准的总传热系数与对流传热系数间的关系为：

$$\frac{1}{K} = \frac{1}{\alpha} + R_{S2} + \frac{bd_2}{\lambda_1 d_m} + R_{S1}\frac{d_2}{d_1} + \frac{d_2}{\alpha_1 d_1} \tag{6-9}$$

式中　d_1，d_2——换热管的外径、内径，m；

　　　d_m——换热管的对数平均直径，m；

　　　b——换热管的壁厚，m；

　　　λ_1——换热管材料的热导率，W/(m·℃)；

　　　R_{S1}，R_{S2}——换热管外侧、内侧的污垢热阻，m²·℃/W。

总传热系数 K 可由式（6-1）求得：

$$K = \frac{Q}{A\Delta t_m} = \frac{m_2 c_{p2}(t_2 - t_1)}{A\Delta t_m} \tag{6-10}$$

用本装置进行实验时，管内空气与管壁间的对流传热系数 α 约为几十到几百 [W/(m²·℃)]，而管外为蒸汽冷凝，冷凝给热系数 α_1 可达 10^4 W/(m²·℃) 左右，因此冷凝传热热阻 $\frac{d_2}{\alpha_1 d_1}$ 可忽略，同时蒸汽冷凝较为清洁，因此换热管外侧的污垢热阻 $R_{S1}\frac{d_2}{d_1}$ 也可忽略。实验中的传热元件材料采用紫铜，热导率 λ_1 为 383.8W/(m·℃)，壁厚为 1.5mm，因此换热管壁的导热热阻 $\frac{bd_2}{\lambda_1 d_m}$ 可忽略。若换热管内侧的污垢热阻 R_{S2} 也忽略不计，则由式（6-9）得

$$\alpha \approx K \tag{6-11}$$

由此可见，被忽略的传热热阻与冷流体侧对流传热热阻相比越小，此法测得的 α 的准确性就越高。

(3) 简易 Wilson 图解法

空气和蒸汽在套管换热器中换热，空气在套管内被套管环隙中的蒸汽加热，当管内空气做充分湍流时，空气侧强制对流传热系数可表示为

$$\alpha = Cu^{0.8} \tag{6-12}$$

将式（6-12）代入式（6-9），得到：

$$\frac{1}{K} = \frac{1}{Cu^{0.8}} + R_{S2} + \frac{bd_2}{\lambda_1 d_m} + R_{S1}\frac{d_2}{d_1} + \frac{d_2}{\alpha_1 d_1} \tag{6-13}$$

依据（2）的分析，式（6-13）右侧后四项在本实验条件下可认为是常数，则由式（6-13）可得：

$$\frac{1}{K} = \frac{1}{Cu^{0.8}} + 常数 \tag{6-14}$$

式（6-14）为 $Y = kX + B$ 线性方程，以 $Y = \frac{1}{K}$，$X = \frac{1}{u^{0.8}}$ 作图，如图 6-2 所示。

由实验线的斜率 $k = \tan\theta$ 得到：

$$\alpha = Cu^{0.8} = \frac{u^{0.8}}{k} \tag{6-15}$$

4. 拟合实验特征数方程式

由实验获取的数据计算出相关特征数 Nu 和 Re 后，在双对数坐标系上，拟合 $Nu \sim Re$ 直线，从而确定拟合方程，得出实验关系式：

$$Nu = ARe^n \tag{6-16}$$

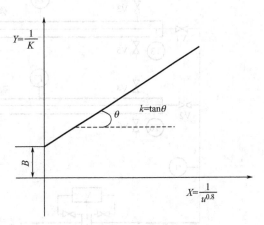

图 6-2 简易 Wilson 图解法作图

式中 Nu——努塞尔数，$Nu = \frac{\alpha d}{\lambda}$，无量纲；

Re——雷诺数，$Re = \frac{du\rho}{\mu}$，无量纲。

5. 传热特征数经验式

对于流体在圆形直管内作强制湍流对流传热时，传热特征数经验式为：

$$Nu = 0.023Re^{0.8}Pr^m \tag{6-17}$$

式中 Pr——普朗特数，$Pr = \frac{c_p\mu}{\lambda}$，无量纲；

λ——定性温度下空气的热导率，W/(m·℃)；

μ——定性温度下空气的黏度，Pa·s。

当流体被加热时 $m = 0.4$，流体被冷却时 $m = 0.3$。

在本实验条件下，考虑 Pr 变化很小，可认为是常数，则式（6-17）改写为：

$$Nu = 0.02Re^{0.8} \tag{6-18}$$

三、实验装置和流程

本实验流程如图 6-3、图 6-4 所示，其中图 6-3 为横管实验装置，图 6-4 为竖管实验装置，横管实验装置虚拟仿真示意图见图 6-5。实验装置由蒸汽发生器、孔板流量计（变送器）、变频器、套管换热器（强化管和普通管）、温度传感器和智能显示仪表等构成。

空气-水蒸气换热流程：来自蒸汽发生器的水蒸气进入套管换热器的壳程，与被风机抽进的空气进行换热交换，不凝气或未冷凝蒸汽通过阀门（V3 和 V4）排出，冷凝水经排出阀（V5 和 V6）排入盛水杯。空气由风机提供，流量通过变频器改变风机转速达到自动控制，空气经孔板流量计进入套管换热器的管程，热交换后从风机出口排出。

注意：普通管和强化管的选取方式，在实验装置上是通过阀门（V1 和 V2）进行切换，仪表柜上通过旋钮进行切换，电脑界面上通过鼠标选取，三者必须统一。

图中符号说明见表 6-1。

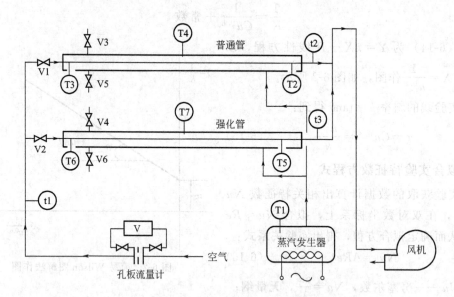

图 6-3 横管对流传热系数测定实验装置流程

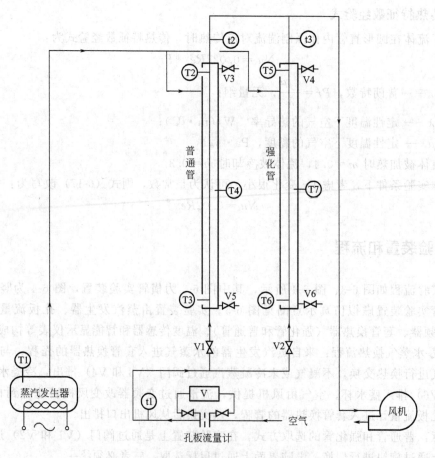

图 6-4 竖管对流传热系数测定实验装置流程

表 6-1 对流传热系数测定实验装置符号说明

符号	名称	单位	备注
V	空气流量	m³/h	
t1	普通管(强化管)空气进口温度	℃	
t2	普通管空气出口温度	℃	紫铜管规格 $\phi 19mm \times 1.5mm$
t3	强化管空气出口温度	℃	即内径为 16mm
T1	蒸汽发生器内的蒸汽温度	℃	有效长度为 1020mm
T2	普通管空气出口端铜管外壁温度	℃	空气流量范围:3~18m³/h
T3	普通管空气进口端铜管外壁温度	℃	V1,V2 为管路空气切换阀门
T4	普通管外蒸汽温度	℃	V3,V4 为不凝气排出阀
T5	强化管空气出口端铜管外壁温度	℃	V5,V6 为冷凝水排出阀
T6	强化管空气进口端铜管外壁温度	℃	
T7	强化管外蒸汽温度	℃	

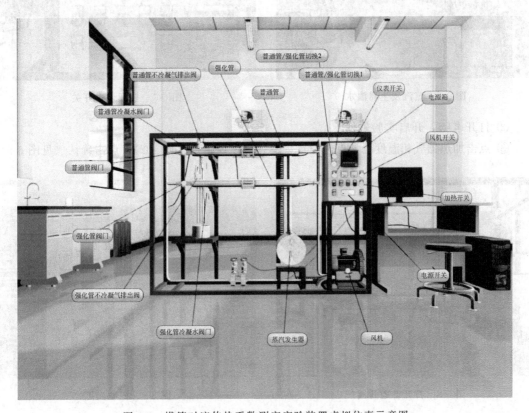

图 6-5 横管对流传热系数测定实验装置虚拟仿真示意图

四、虚拟仿真实验操作步骤

1. 检查实验装置,开始实验

① 实验前先检查仪表、风机、蒸汽发生器及测温点是否正常。

扫一扫动画
对流传热系数的测定实验

② 打开蒸汽发生器盖子，用烧杯向水箱注入水。

③ 同时注意查看液面指示计，至少注入 4/5 的水量（见图 6-6），然后盖上水箱盖子。

④ 打开电源开关（见图 6-7）、仪表开关、加热开关。

2. 普通管实验

① 打开普通管阀门（见图 6-8），开启不冷凝气排出阀至合适开度。

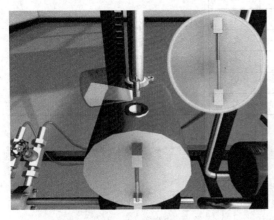

图 6-6　蒸汽发生器灌水

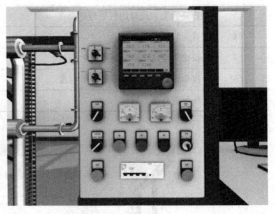

图 6-7　打开仪表开关

② 打开电脑，开启实验软件。

③ 点击加热器气相温度处，点击设置，输入温度设定值（如102），点击确认（见图 6-9）。

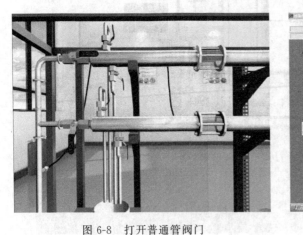

图 6-8　打开普通管阀门

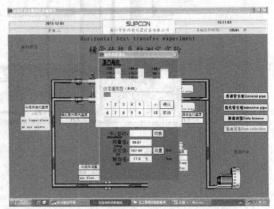

图 6-9　控制温度

④ 等有大量气体冒出，伴随蒸汽排出时，表明不凝气排放干净。

⑤ 将普通管冷凝水排出阀打开至在冷凝水排出的前提下有微量蒸汽排出的开度（见图 6-10）。

⑥ 打开控制柜上风机开关，点击软件冷流体流量处，设置流体（空气）流量（见图 6-11）。

⑦ 将仪表柜和软件都切换为普通管实验。

⑧ 待数据稳定之后，采集普通管数据。

⑨ 合理分配调节流量，记录数据。

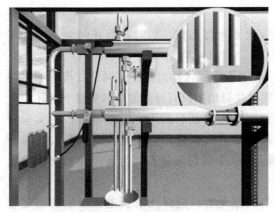

图 6-10　产生蒸汽

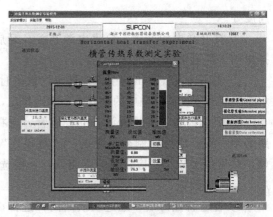

图 6-11　流量设定

3. 加强管实验

① 关闭普通管阀门及冷凝水排出阀。

② 打开强化管阀门（见图 6-12）。

③ 打开并调节强化管冷凝水排出阀至适合位置，在保证冷凝水排出的前提下有微量蒸汽排出。

④ 将仪表柜和软件都切换为强化管实验。

⑤ 点击软件中冷流体流量处，设置冷流体（空气）流量。

⑥ 等待流体流量和热交换稳定后采集数据（见图 6-13）。

⑦ 合理分配调节流量，记录数据。

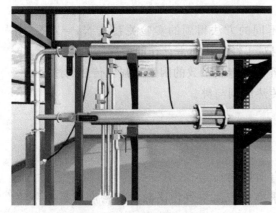

图 6-12　打开强化管阀门

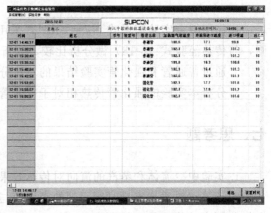

图 6-13　数据采集

4. 实验结束，整理仪器

① 关闭蒸汽发生器，实验切换归位，关闭仪表。

② 待蒸汽发生器内温度下降至 95℃ 以下后，再关闭风机电源，最后关闭总开关。

③ 关闭阀门，最后关闭电脑。

五、实验方法及步骤

1. 实验操作步骤

① 检查仪表、风机、蒸汽发生器及测温点是否正常，将蒸汽发生器灌水至液位计的 4/5

高度处。

②打开总电源开关、仪表电源开关，开启蒸汽发生器加热。同时，全开两个不冷凝气排出阀，控制温度在100℃左右。

③当不冷凝气排出阀中有大量蒸汽冒出时，关闭不冷凝气排出阀；再调节好换热器冷凝水排出阀开度，使环隙中的冷凝水不断地排出，同时使排出的没有冷凝的蒸汽量越小越好。

④启动风机，先选择设备的普通管，选择仪表及电脑显示的测定管路与设备的实验管路必须一致。通过控制软件上的"流量设定"来调节空气流量，空气流量应从最大流量开始，然后依次减小。应合理分配流量，一般选择3~5个实验流量点，待流量和热交换稳定后，再采集数据。

⑤普通管测好后，通过切换阀门，选择设备的强化管，选择仪表及电脑显示的测定管路与设备的实验管路要一致，数据测定方法同步骤④。

⑥实验结束时，先关闭蒸汽发生器电源，待蒸汽发生器内温度下降至95℃以下后，再关闭风机电源，关闭总电源，做好清洁工作。

2. 注意事项

①开始实验时，必须先检查蒸汽发生器中的水位是否正常，发现水位过低，应及时补充水。

②调节流量后，至少稳定30min后读取实验数据。

③实验中保持上升蒸汽的稳定，且蒸汽放空口一直有蒸汽逸出。

六、实验报告

1. 作图拟合实验关系式 $Nu = ARe^n$，采用不同的方法得到的实验关系式 $Nu = ARe^n$ 不同，与传热经验式（6-18）做分析比较，分析相对于经验式而言存在的差异。
2. 通过普通管和强化管实验结果的对比分析影响传热系数的因素和强化传热的途径。
3. 分析本实验的强化管传热相对于普通管传热的利与弊。

七、思考题

1. 实验中冷流体和蒸汽的流向对传热效果有何影响？
2. 计算空气质量流量时所用到的密度值与求换热管内空气雷诺数时的密度值是否一致？它们分别表示什么位置的密度？
3. 实验过程中，冷凝水不及时排走，会产生什么影响？如何及时排走冷凝水？如果采用不同压强的蒸汽进行实验，对 α 关联式有何影响？
4. 就本实验过程而言，为提高总传热系数 K，可采用哪些有效的方法，其中最有效的方法是什么？
5. 本实验主要依据哪个参数的稳定作为实验采集数据的前提，为什么？

第七章

过滤实验——恒压过滤、真空过滤

一、实验目的和要求

1. 了解和掌握板框压滤以及真空过滤的设备构造和操作方法。
2. 测定恒压过滤和真空过滤的过滤常数 K、q_e，虚拟过滤时间 τ_e 及压缩性指数 s。
3. 了解过滤压力对过滤速率的影响。

二、实验原理

过滤是以某种多孔物质为介质来处理悬浮液以达到固、液分离的一种操作过程，即在外力的作用下，悬浮液中的液体通过固体颗粒层（即滤渣层）及多孔介质的孔道而固体颗粒被截留下来形成滤渣层，从而实现固、液分离。因此，过滤操作本质上是流体通过固体颗粒层的流动，而这个固体颗粒层（滤渣层）的厚度随着过滤的进行不断增加，故在恒压过滤操作中，过滤速率不断降低。

过滤设备的生产能力用过滤速率来表示，过滤速率是单位时间内通过单位过滤面积的滤液体积。影响过滤速率的基本因素为：过滤压力（压强差）Δp、滤渣厚度 L、悬浮液的性质、悬浮液温度及液体的黏度等。

过滤速率：

$$u = \frac{dV}{Ad\tau} = \frac{dq}{d\tau} = \frac{A\Delta p^{(1-s)}}{\mu r C(V+V_e)} = \frac{A\Delta p^{(1-s)}}{\mu r'C'(V+V_e)} \tag{7-1}$$

式中　u——过滤速率，m/s；

　　　V——通过过滤介质的滤液体积量，m³；

　　　A——过滤面积，m²；

　　　τ——过滤时间，s；

　　　q——通过单位面积过滤介质的滤液量，m³/m²；

　　　Δp——过滤压力（表压），Pa；

　　　s——滤渣压缩性指数；

　　　μ——滤液的黏度，Pa·s；

　　　r——滤渣比阻，1/m²；

　　　C——单位滤液体积的滤渣体积，m³/m³；

V_e——过滤介质的当量滤液体积，m^3；

r'——滤渣比阻，m/kg；

C'——单位滤液体积的滤渣质量，kg/m^3。

在恒压下过滤操作时，Δp 为常数，在工作温度一定下，μ，r，C 也为常数，则令：

$$K = \frac{2\Delta p^{(1-s)}}{\mu r C} \tag{7-2}$$

于是式（7-1）可改写为：

$$\frac{dV}{d\tau} = \frac{KA^2}{2(V+V_e)} \tag{7-3}$$

式中 K——过滤常数，由物料特性及过滤压差所决定，m^2/s。

将式（7-3）分离变量并积分，整理得：

$$\int_{V_e}^{V+V_e} (V+V_e) d(V+V_e) = \frac{1}{2} KA^2 \int_0^\tau d\tau \tag{7-4}$$

即：

$$V^2 + 2VV_e = KA^2 \tau \tag{7-5}$$

将式（7-4）的积分极限改为从 0 到 V_e 和从 0 到 τ_e，则：

$$V_e^2 = KA^2 \tau_e \tag{7-6}$$

式（7-5）和式（7-6）相加可得：

$$(V+V_e)^2 = KA^2(\tau+\tau_e) \tag{7-7}$$

式中 τ_e——虚拟过滤时间，相当于滤出滤液量 V_e 所需时间，s。

将式（7-7）微分，得：

$$2(V+V_e) dV = KA^2 d\tau \tag{7-8}$$

即：

$$\frac{\Delta \tau}{\Delta q} = \frac{2}{K}\bar{q} + \frac{2}{K}q_e \tag{7-9}$$

式中 Δq——每次测定的单位过滤面积滤液体积（在实验中一般等量分配），m^3/m^2；

$\Delta \tau$——每次测定的滤液体积 Δq 所对应的时间，s；

$\bar{q}$——相邻两个 q 值的平均值，m^3/m^2。

以 $\Delta \tau/\Delta q$ 为纵坐标，$\bar{q}$ 为横坐标将式（7-9）绘制成一直线，即恒压过滤曲线，可得：

斜率：

$$S = \frac{2}{K}$$

截距：

$$I = \frac{2}{K} q_e$$

则

$$K = \frac{2}{S}, \ m^2/s$$

$$q_e = \frac{KI}{2} = \frac{I}{S}, \text{ m}^3$$

$$\tau_e = \frac{q_e^2}{K} = \frac{I^2}{KS^2}, \text{ s}$$

改变过滤压差 Δp，可测得不同的 K 值，由 K 的定义式（7-2）两边取对数得：

$$\lg K = (1-s)\lg(\Delta p) + B \tag{7-10}$$

在实验压差范围内，若 B 为常数，则 $\lg K \sim \lg(\Delta p)$ 的关系在直角坐标上应是一条直线，斜率为 $(1-s)$，可得滤饼压缩性指数 s。

三、实验装置和流程

1. 恒压过滤

本装置压滤机采用不锈钢切向型多层压滤机，由板框压滤机、配料槽、料浆泵、恒压料罐、恒压储水罐、计量桶及空气压缩机（简称空压机）等组成，其流程如图 7-1 所示。虚拟仿真示意图见图 7-2。

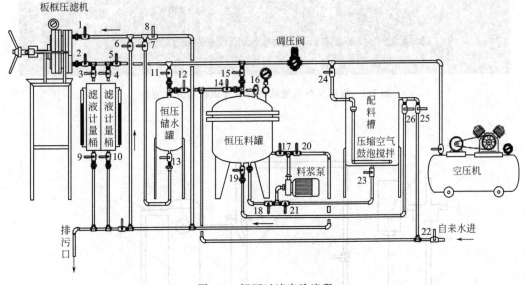

图 7-1　恒压过滤实验流程

CaCO₃悬浮液在配料槽内制成一定浓度后由料浆泵送入恒压料罐，并用料浆泵对恒压料罐内料液进行不断的循环搅拌，并用压缩空气（恒压）将滤浆压入板框压滤机进行压滤，滤液由计量桶对其进行计时、计量。

2. 真空过滤

称取一定量的 CaCO₃，在恒温滤浆槽内配制一定浓度的 CaCO₃ 悬浮液，用电动搅拌器进行均匀搅拌，启动真空泵，使系统内形成真空并达指定值，然后打开过滤漏斗上的球阀使浆液在压差推动下流入过滤漏斗，经过滤后的清液流入计量桶计量。实验流程及仿真示意图见图 7-3、图 7-4。

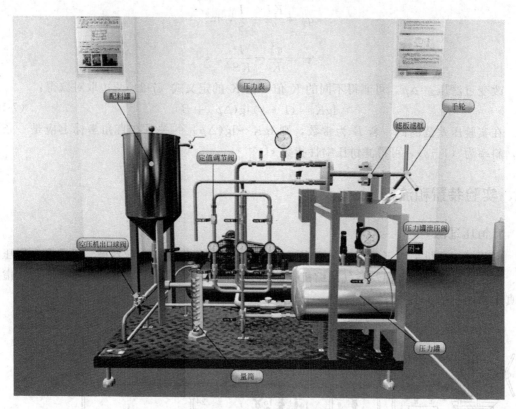

图 7-2 板框压滤机过滤实验装置虚拟仿真示意图

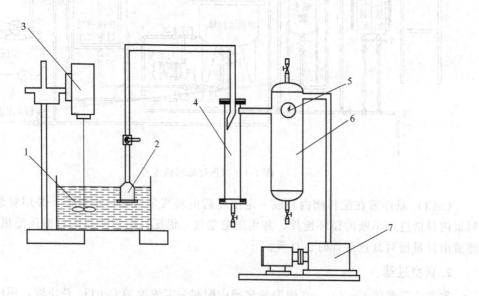

图 7-3 真空过滤实验流程示意图

1—恒温滤浆槽；2—过滤漏斗；3—搅拌电机；
4—计量桶；5—真空压力表；6—缓冲罐；
7—真空泵

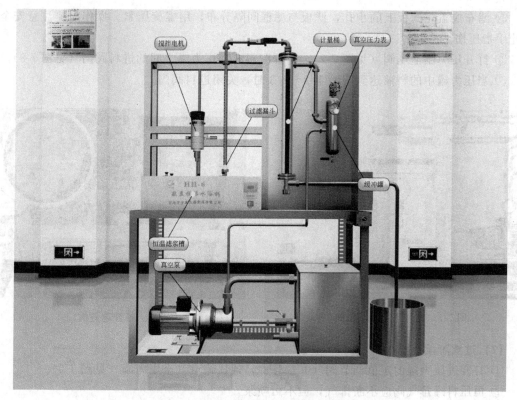

图 7-4 真空过滤实验装置虚拟仿真示意图

四、虚拟仿真实验操作步骤

1. 恒压过滤

(1) 准备阶段

① 稍稍打开空压机出口阀门（见图 7-5）。
② 打开配料罐出料口。
③ 缓慢调节空压机出口阀门。
④ 安装板框（见图 7-6），正确安装滤板、滤框及滤布。

扫一扫动画
恒压过滤

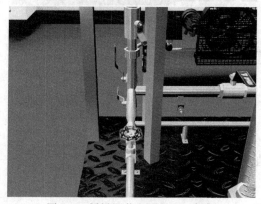

图 7-5 缓慢调节空压机出口阀门

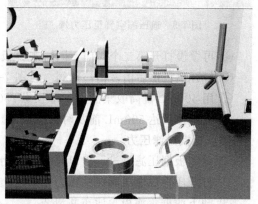

图 7-6 安装板框

第七章 过滤实验——恒压过滤、真空过滤

⑤ 滤布应盖住滤板上的小孔，滤板与滤框间隔分布，用螺旋压紧。操作时要注意安全，防止手指压伤。

⑥ 打开压力罐泄压阀（见图7-7），打开配料罐和压力罐之间的进料阀门（见图7-8）。

⑦ 当压力罐中的料液达到视镜1/2～2/3时，关闭进料阀门。

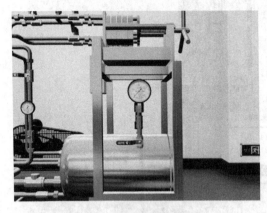
图7-7 打开压力罐泄压阀

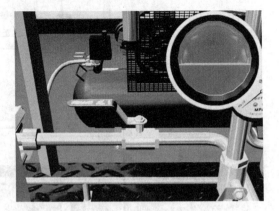
图7-8 打开进料阀门

(2) 过滤阶段

① 打开进压力罐的定值调节阀（0.1MPa），通压缩空气至压力罐（见图7-9）。

② 恒压料罐排气阀应不断排气，但不可喷浆。

③ 先打开进板框前料液进口的一个阀门（见图7-10）。

图7-9 通压缩空气至压力罐

图7-10 打开进板框前料液进口阀门

④ 再缓慢打开第二个阀门，同时观察板框有没有出现喷浆。

⑤ 打开出板框后清液出口球阀（见图7-11），清液出口流出滤液。

⑥ 用500mL量筒收集滤液，每收集250mL滤液记录相应的过滤时间（见图7-12）。

⑦ 在液体到达500mL前及时更换量筒。

(3) 更换过滤压力

① 关闭板框压滤机进、出口阀门，关闭定值调节阀。

② 打开压力罐泄压阀，使压力罐泄压。

③ 卸下滤框、滤板、滤布并清洁。

图 7-11 打开出板框后清液出口球阀

图 7-12 记录相应的过滤时间

④ 开始下一组实验前，安装板框，将泄压阀关小。

(4) 过滤阶段

① 打开进压力罐的第二路定值调节阀门（0.2MPa），通压缩空气至压力罐。
② 先打开进板框前料液进口的一个阀门。
③ 再缓慢打开第二个阀门，同时观察板框有没有出现喷浆。
④ 打开出板框后清液出口球阀，清液出口流出滤液，用 500mL 量筒收集滤液。
⑤ 每收集 250mL 滤液记录相应的过滤时间，在液体到达 500mL 前需及时更换量筒。
⑥ 改变过滤压力为 0.25MPa，重复过滤操作。

(5) 结束阶段

① 关闭板框压滤机进、出口阀门。
② 关闭定值调节阀。
③ 打开压力罐泄压阀，使压力罐泄压。
④ 卸下滤框、滤板、滤布并清洗。
⑤ 关闭配料罐出料口，关闭空压机出口阀门。

2. 真空过滤

(1) 实验前准备

① 检查各电器设备是否工作正常。
② 检查计量桶刻度是否标记正确。
③ 接通电源。

(2) 开始操作

① 开启搅拌电机进行搅拌（见图 7-13），转速设置在 0.08~0.10kr/min。
② 打开缓冲罐放空阀（见图 7-14）。
③ 开启真空泵（见图 7-15），调节缓冲罐真空度至 0.06MPa。
④ 注意计量桶中留有一定的液体作为零液位。
⑤ 打开过滤漏斗上的球阀（见图 7-16）。
⑥ 恒温滤浆槽内浆液在压强差的推动下通过过滤漏斗过滤，清液流入计量桶内。
⑦ 滤液高度每增加 2cm，用秒表读取相应时间并记录，测量 10 个读数（见图 7-17）。

扫一扫动画
真空过滤

图 7-13　开启搅拌电机

图 7-14　打开缓冲罐放空阀

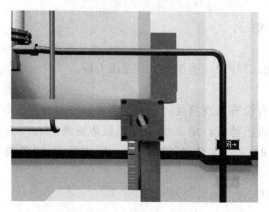

图 7-15　开启真空泵

图 7-16　打开过滤漏斗上的球阀

(3) 改变条件

① 停真空泵，关闭过滤漏斗上的球阀。

② 打开放液阀排出计量桶内的清液（见图 7-18）。

图 7-17　读取时间并记录

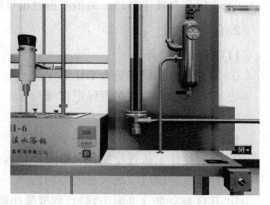

图 7-18　打开放液阀

③ 保留一定的零液位，高度和第一次实验选取的高度相同。

④ 卸下过滤漏斗、滤布进行清洗。

⑤ 开启真空泵。

⑥ 调缓冲罐真空度至 0.05MPa，重复操作。
⑦ 测完后，改变真空度为 0.04MPa，重复操作。
(4) 实验结束，整理实验仪器
① 停真空泵，关闭过滤漏斗上的球阀。
② 关闭搅拌电机，切断总电源。
③ 打开计量桶排液阀，排净计量桶内清液，清洗实验设备。

五、实验方法及步骤

1. 恒压过滤

(1) 实验操作步骤

① 配料：开阀 25 将清水放入配料槽约 25kg 后关阀 25，然后称取 $CaCO_3$ 粉末 1.30kg 倒入配料槽内。

② 启动空气压缩机（空压机限定压力可设置到 0.2MPa），开阀 24 用压缩空气对料液进行搅拌，开阀 23、21、17，并启动泵，将滤浆全部打入恒压料罐。然后关阀 23、21，再开阀 18，使恒压料罐内浆液由料浆泵进行循环搅拌。

③ 开阀 12，同时开阀 13，将清水注入恒压储水罐内，一定时间后关阀 12、13。

④ 调节恒压阀压力指数，一般调至 0.05～0.10MPa，然后开阀 11、15。

⑤ 开阀 1、2、6，滤浆随压缩空气压入压滤机，开阀 3 或阀 4（计量时只能开其中一只），滤液就进入计量桶，并对滤液量和其对应时间进行测量，当压滤机进口压力表读数增大至与恒压料罐压力相近时，结束本次测量，开阀 9 或阀 10 将滤液排出。

⑥ 关阀 6，开阀 7，用清水对滤渣进行洗涤。

⑦ 关阀 3（或阀 4）及阀 7，开阀 5、阀 8，用压缩空气对滤渣吹气，有利于滤渣的脱落，然后关阀 5、阀 8，卸滤渣并洗净滤布。若恒压料罐内有余料，可开阀 26 或阀 19，用空气将余料压入配料槽内或排出，最后关阀 11、阀 15，关空气压缩机。

⑧ 实验结束后，还应对设备进行清洗。

(2) 注意事项

① 启动空气压缩机之前要熟悉整个流程气路走向。
② 操作压力不宜过大，滤布要铺平整，切忌压力过大导致滤浆从板框中喷出。
③ 正确操作排气阀和压力调节阀。

2. 真空过滤

(1) 实验操作步骤

① 实验前操作准备：检查管道连接口有没有漏气，抱箍有没有松动。检查各电器设备是否工作正常。检查计量桶刻度是否标记正确。确认真空泵开关、恒温滤浆槽加热开关处于关闭状态。接通电源（把电源插头插到插座上）。关闭设备上的所有阀门。

② 配料：在恒温滤浆槽内配制含 $CaCO_3$ 5%～10%（质量分数）的水悬浮液，碳酸钙事先由天平称重，恒温滤浆槽的尺寸为 490×325×115(mm)。

③ 搅拌：开启搅拌电机进行搅拌，控制好搅拌转速，一般转速设置在 0.08～0.10kr/min。

④ 系统抽真空：先打开缓冲罐放空阀，再开启真空泵，然后通过调节缓冲罐放空阀开度，使系统内真空度达到指定值（0.06MPa、0.05MPa、0.04MPa）。

⑤ 真空过滤：打开过滤漏斗上的球阀，使恒温滤浆槽内浆液在压强差的推动下通过过滤漏斗过滤，清液流入计量桶内。真空压力表指示过滤真空度。在不同真空度下，记录10~12个读数即可停止实验。

⑥ 实验结束操作：停真空泵，关闭搅拌电机，切断总电源，打开计量桶排液阀，排净计量桶内清液。

(2) 注意事项

① 真空度一般在 0.04~0.06MPa，如果真空度过小则放空阀开度过大，真空泵在畅通大气下工作，此时工作时间一般在 3min 以内；如果过大则导致管道连接处承载压力过大，使连接处有气体进入。真空泵长时间使用后，如果真空度过大会导致泵出口有大量白烟冒出，此时应减小真空度或停泵冷却。

② 计量桶内液位不得超过进缓冲罐管口的高度，防止液体吸入真空泵。

六、实验报告

1. 由真空过滤实验数据求过滤常数 K、q_e 及 τ_e。
2. 比较不同压差下的 K、q_e、τ_e 值，讨论压差变化对以上参数数值的影响。
3. 在直角坐标系上绘制 $\Delta \tau / \Delta q \sim \bar{q}$ 关系曲线及 $\lg K \sim \lg \Delta p$ 关系曲线，求出 s。
4. 实验结果分析与讨论。

七、思考题

1. 板框压滤机的优缺点是什么？适用于什么场合？
2. 板框压滤机的操作分哪几个阶段？
3. 为什么过滤开始时，滤液常常有点浑浊，而过段时间后才变清？
4. 影响过滤速率的主要因素有哪些？在某一恒压下测得 K、q_e、τ_e 值后，若将过滤压强提高一倍，问上述三个值将有何变化？

第八章

填料塔吸收过程实验

一、实验目的和要求

1. 了解填料吸收塔的基本构造并熟悉吸收塔操作的基本流程。
2. 观察填料吸收塔的液泛现象。
3. 测定填料层压降 Δp 与空塔气速 u 的关系曲线，测定泛点空塔气速。
4. 测定含氨空气-水系统的体积吸收系数 $K_Y a$。

二、实验原理

气体吸收是典型的传质过程之一。

1. 填料层压力降 Δp 与空塔气速 u 的关系

填料塔的压力降与空塔气速是填料塔设计与操作中重要的流体力学参数。

气体通过干填料层时（喷淋密度 $L_0 = 0$），其压力降 Δp 随空塔气速 u 的变化如图 8-1 中直线 0 所示，此直线斜率约为 1.8，与气体以湍流方式通过管道时 Δp 与 u 的关系相仿。由图 8-1 可知，在一定喷淋密度下（L_1，L_2，L_3），当气速在 L 点以下时，由于持液使得填料层空隙率减小，故其压降高于相同空塔气速下的干塔压降，又由于此时气体对液膜的流动无明显影响，持液量不随气速变化，故其 $\Delta p \sim u$ 关系与干填料基本平行。

在一定喷淋密度下，气速增大至一定程度时，随气速增大，液膜增厚，即出现"拦液状态"（如图 8-1 中 L 点以上），此时气体

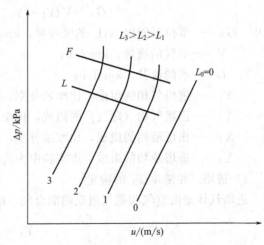

图 8-1 填料塔压降 Δp 与空塔气速 u 的关系

通过填料层的流动阻力剧增；若气速继续加大，上升气流对液体所产生的曳力使得流体向下流动严重受阻，集聚的液体从填料表面扩展到整个填料层，谓之"液泛状态"（如图 8-1 中 F 点以上），此时气体的流动阻力急剧增加，填料层的压降几乎垂直上升。图 8-1 中 F 点即

为泛点，与之相对应的气速 u 称为泛点气速。

如填料塔内空塔气速超过液泛点，则全塔压降急剧上升，使得塔的操作过程极不稳定，因此通常实际操作气速取泛点气速的 60%～80%。

(1) 空塔气速 u

塔内气体的流速以其体积流量与塔截面积之比来表示，称为空塔气速 u。

$$u = \frac{V'}{\Omega} \tag{8-1}$$

式中　u——空塔气速，m/s；

　　　V'——塔内气体体积流量，m^3/s；

　　　Ω——塔截面积，m^2。

实验中气体流量（空气和氨气）由转子流量计测量。由于实验气体条件与转子流量计标定条件不相同，故转子流量计的读数值必须进行校正。气体转子流量计校正式见式（8-2）

$$Q_S = Q_N \sqrt{\frac{p_N T_S}{p_S T_N}} \tag{8-2}$$

式中　Q_S——气体实际流量，m^3/h 或 L/h；

　　　Q_N——气体流量计读数，m^3/h 或 L/h；

p_N，T_N——标定气体的标定状况，$p_N = 1.013 \times 10^5 Pa$，$T_N = 293K$；

p_S，T_S——被测气体的绝对压强，热力学温度，Pa，K。

(2) 填料层压降 Δp

填料层压降 Δp 直接可由 U 形压差计 14 读取。

2. 体积吸收系数 $K_Y a$ 的测定

(1) NH_3 的吸收量 G_A

由物料衡算求得：

$$G_A = V(Y_1 - Y_2) = L(X_1 - X_2) \tag{8-3}$$

式中　G_A——单位时间内 NH_3 的吸收量，kmol/h；

　　　V——空气的流量，kmol/h；

　　　L——水的流量，kmol/h；

　　　Y_1——进塔气相的组成，比摩尔分率，kmol(A)/kmol(B)；

　　　Y_2——出塔气相（尾气）的组成，比摩尔分率，kmol(A)/kmol(B)；

　　　X_1——出塔液相的组成，比摩尔分率，kmol(A)/kmol(B)；

　　　X_2——进塔液相的组成，本实验中为清水吸收，$X_2 = 0$。

1) 进塔气相浓度 Y_1 的确定

进塔气体是由空气和氨气组成的混合气，故进塔气相浓度 Y_1

$$Y_1 = \frac{V_A}{V} \tag{8-4}$$

式中　V_A——进塔氨气的摩尔流量，kmol/h；

　　　V——进塔空气的摩尔流量，kmol/h。

2) 出塔气相（尾气）组成 Y_2 的测定方法

用移液管移取 V_a mL 浓度为 M_a(mol/L) 的标准 H_2SO_4 溶液置于洗气瓶中，加入适量去离子水和 3～5 滴溴百里酚蓝，如装置图（图 8-2）所示连接在取样尾气管线上。当吸收塔操作稳定时,改变三通旋塞方向,使部分塔顶尾气通过洗气瓶,尾气中的氨气被 H_2SO_4 中和吸收,从洗气瓶逸出的空气进入湿式流量计计量。当瓶内溶液颜色由黄色(酸性)变至绿色(中性)时,表明尾气吸收至终点,迅速改变三通旋塞方向,即切断气体进入洗气瓶,否则溶液颜色将变成蓝色(碱性)。

$$Y_2 = \frac{n_{NH_3}}{n_{air}} \tag{8-5}$$

式中 n_{NH_3}——氨气的摩尔数,mol;

n_{air}——空气的摩尔数,mol。

氨气与硫酸发生中和反应,由硫酸量得到

$$n_{NH_3} = 2M_a V_a \times 10^{-3} \tag{8-6}$$

式中 V_a——标准 H_2SO_4 溶液体积,mL;

M_a——标准 H_2SO_4 溶液浓度,mol/L。

由湿式流量计读数差值获得空气体积量 V_0,并测其温度 T_0。

$$n_{air} = \frac{p_0 V_0}{RT_0} \tag{8-7}$$

式中 p_0——通过湿式流量计的空气压力(可取室内大气压),Pa;

V_0——通过湿式流量计的空气体积,L;

T_0——通过湿式流量计的空气温度,K;

R——气体常数,$R=8314$N·m/(mol·K)。

(2) 相平衡常数 m

本实验为低浓度气体吸收,相平衡关系遵循亨利定律的物系,气液平衡关系式为:

$$y^* = mx \tag{8-8}$$

相平衡常数 m 与系统总压 p 和亨利系数 E 的关系为:

$$m = \frac{E}{p} \tag{8-9}$$

式中 E——亨利系数,Pa;

p——系统总压(实验中取塔内平均压力),Pa。

根据实验中所测得的塔顶表压、塔顶塔底压差即可求得塔内平均压力 p。

本实验体系亨利系数 E 与温度 T 的关系为:

$$\lg E = 11.468 - 1922/T \tag{8-10}$$

式中 T——吸收温度(实验中取塔底液相温度),K。

(3) 气相对数平均浓度差 ΔY_m

$$\Delta Y_m = \frac{\Delta Y_1 - \Delta Y_2}{\ln \frac{\Delta Y_1}{\Delta Y_2}} \tag{8-11}$$

其中 $\Delta Y_1 = Y_1 - Y_1^* = Y_1 - mX_1$

$\Delta Y_2 = Y_2 - Y_2^* = Y_2 - mX_2$

式中 Y_1^*,Y_2^*——与液相浓度 X_1、X_2 相对应的气相平衡浓度,kmol(A)/kmol(B)。

(4) 气相体积吸收系数 $K_Y a$

气相体积吸收系数 $K_Y a$ 是反映填料吸收塔性能的主要参数之一,其值也是设计填料塔的重要依据。本实验属于低浓度气体吸收，近似取 $Y \approx y$、$X \approx x$。
则

$$K_Y a = \frac{G_A}{\Omega h \Delta Y_m} \tag{8-12}$$

式中 $K_Y a$——气相体积吸收系数，$kmol/(m^3 \cdot h)$；

a——单位体积填料层所提供的有效传质面积，m^2/m^3；

G_A——单位时间内 NH_3 的吸收量，$kmol/h$；

Ω——塔截面积，m^2；

h——填料层高度，m；

ΔY_m——气相传质推动力，气相对数平均浓度差。

三、实验装置和流程

① 本实验装置的流程及虚拟仿真示意图如图 8-2、图 8-3 所示。主体设备是内径为 70mm 的吸收塔，塔内装 10mm×9mm×1mm 的陶瓷拉西环填料。

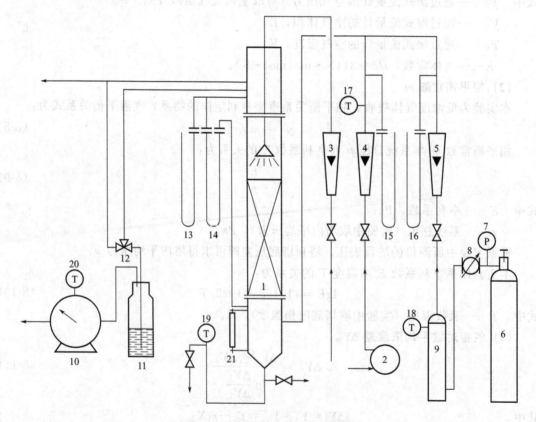

图 8-2 填料塔吸收过程实验装置流程

1—填料吸收塔；2—旋涡气泵（风机）；3—流量计（水）；4—流量计（空气）；5—流量计（氨气）；
6—液氨钢瓶；7—氨压力表；8—氨气减压阀；9—氨气缓冲罐；10—湿式流量计；
11—洗气瓶；12—三通旋塞；13~16—U形压差计；
17~20—温度计；21—液面计

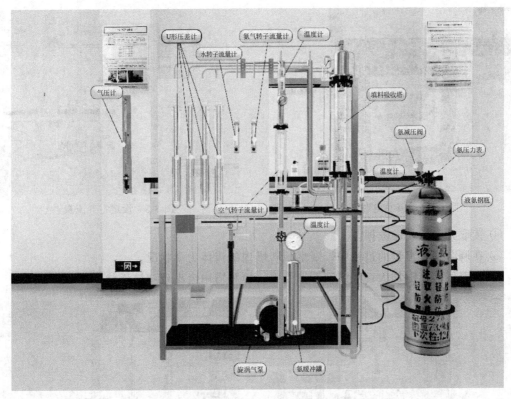

图 8-3 填料塔吸收过程实验装置虚拟仿真示意图

② 物料体系是水-空气-氨气，空气由漩涡气泵送出后，用转子流量计 4 测其流量；氨气由液氨钢瓶供应，经转子流量计 5 测量其流量；吸收剂（水）由转子流量计 3 测其流量。水从塔顶喷淋至填料层，与自下而上的含氨空气进行吸收过程，溶液由塔底经液封管流出，塔底有液相取样口，经吸收后的尾气由塔顶排出，自塔顶引出适量尾气，用化学分析法对其进行浓度分析。

四、虚拟仿真实验操作步骤

1. 在水喷淋密度为 0 的情况下，进行实验

① 读取福廷气压计读数，作为本次实验的大气压值。
② 打开总电源开关（见图 8-4），启动风机。
③ 开启空气管道阀门，将空气通入填料吸收塔。
④ 调节空气流量，使空气流量达到设定值（见图 8-5）。
⑤ 记录空气表压、塔顶表压、塔顶塔底压差、空气温度。
⑥ 调节空气流量（分配流量），并记录调整空气流量后的实验数据。

扫一扫动画
填料塔吸收过程实验

2. 将水喷淋密度调至 30L/h，进行实验

① 开启水调节阀。
② 调节水转子流量计，使示数稳定在设定值，以 30L/h 为例（见图 8-6）。
③ 调整空气阀门开闭状态，调节空气流量（分配流量），使空气流量达到设定值。

第八章 填料塔吸收过程实验

图 8-4 打开总电源开关

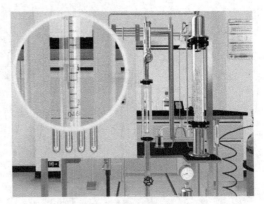

图 8-5 设定空气流量

④ 读取并记录实验数据。
⑤ 在调节空气流量的过程中,观察填料塔内的流体力学状况。
⑥ 随着气体流速加大,喷淋液下流受阻,填料塔中会出现"液泛状态"(见图 8-7)。

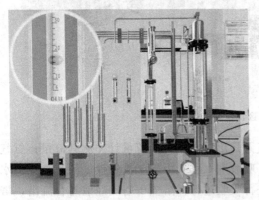

图 8-6 设定水流量

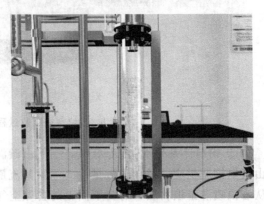

图 8-7 液泛现象

3. 组装尾气吸收装置

① 通入氨气之前需要组装好尾气吸收装置。
② 用移液管移取标准硫酸溶液置于洗气瓶中(见图 8-8)。
③ 加入 2~3 滴溴百里酚蓝。
④ 加入去离子水,加至洗气瓶的 2/3 处(见图 8-9)。

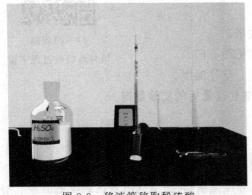

图 8-8 移液管移取稀硫酸

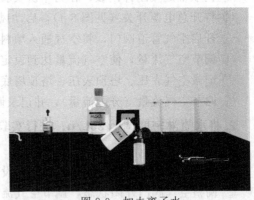

图 8-9 加去离子水

⑤ 将洗气瓶进口端连接在塔顶尾气出口管线上，出口端连接到湿式流量计。
⑥ 记录湿式流量计的初始数据。

4. 测定含氨空气-水系统的体积吸收系数

① 调节空气阀门，使空气转子流量计示数稳定在设定值（如 $8m^3/h$）。
② 调节水阀门，使水转子流量计示数稳定在设定值（如 30L/h）。
③ 打开液氨钢瓶总阀，缓慢调节钢瓶的减压阀。
④ 需根据空气流量估算氨气流量（进塔气相浓度约为 3%），调节氨气转子流量计，使其稳定（见图 8-10）。
⑤ 操作稳定后，开启三通旋塞，将旋塞旋转 180°，使尾气通入洗气瓶进行尾气组成分析。
⑥ 当洗气瓶内溶液颜色由黄色变至绿色，说明尾气吸收至终点，需立即关闭三通旋塞，防止尾气过量（见图 8-11）。
⑦ 若通入的尾气不慎过量，洗气瓶内溶液变为蓝色，则需要重新进行尾气分析。

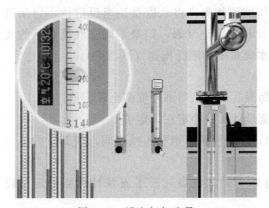

图 8-10 设定氨气流量

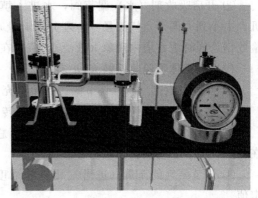

图 8-11 塔顶气相分析

5. 读取并记录数据

① 记录尾气吸收至终点时湿式流量计数据（见图 8-12）。
② 记录通过湿式流量计的空气温度。
③ 记录氨气温度、氨气表压、空气表压、塔顶表压、塔顶塔底压差、空气温度、塔底液温（见图 8-13）。
④ 改变实验条件，重复进行以上操作。

图 8-12 读取湿式流量计数据

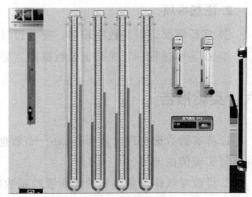

图 8-13 读取表压

第八章 填料塔吸收过程实验

6. 实验结束，关闭实验设备

① 关闭氨气转子流量计。
② 关闭水转子流量计。
③ 关闭空气转子流量计。
④ 关闭氨气阀门、水阀门，关闭风机电源开关，清理实验仪器。

五、实验方法及步骤

1. 实验操作步骤

(1) 测定 $\Delta p \sim u$ 实验

① 在水喷淋密度为零的情况下，启动风机，由小到大调节空气流量（合理分配流量），设计实验记录表，记录不同空气流量下的实验数据。

② 开启吸收剂（水）调节阀，水进入填料塔润湿物料。将水流量调至设定值，启动风机，由小到大调节空气流量（合理分配流量），设计实验记录表，记录不同空气流量下的实验数据，同时观察塔内现象（液泛），获得在该水流量下的液泛气速（液泛点）。

(2) 测定体积吸收系数 $K_Y a$ 实验

① 开启吸收剂（水）调节阀，当填料充分润湿后，将水喷淋密度调至设定值。
② 启动风机，调节空气流量，使空气转子流量计示数稳定在设定值。
③（在教师指导下操作）打开液氨钢瓶总阀，并缓慢调节钢瓶的减压阀，使其压力稳定在 0.1~0.2MPa 左右。
④ 调节氨气的流量（进塔气相氨的浓度约为 3%），使氨气转子流量计读数稳定在设定值。
⑤ 配制分析塔顶气相组成的酸性吸收液，准确移取一定量的吸收液至洗气瓶内。
⑥ 待系统稳定后，开启三通旋塞，使得部分塔顶气体通入洗气瓶，进行塔顶气相组成分析，设计实验记录表，分工合作，记录流量、温度、压差及塔顶气相组成分析等数据。
⑦ 改变操作条件（水流量），重复步骤⑥。
⑧ 再次改变操作条件（空气和氨气流量），重复步骤⑥。
⑨ 实验完毕，首先关闭氨气转子流量计，再关闭空气和水转子流量计，关闭风机电源开关，清理实验仪器。

2. 注意事项

① 实验需要记录的数据较多，需要设计完整的数据记录表，分工合作完成实验。
② 在实验过程中，尤其记录数据时，要确保空气、氨气和水流量的稳定。

六、实验报告

1. 将实验数据整理计算得到 $\Delta p \sim u$ 数据表，由数据表作图 $\Delta p \sim u$，得出液泛点，比较该值与实验值的差异。

2. 将改变气体流量得到的体积吸收系数的变化与改变液量得到的体积吸收系数的变化作比较，从传质理论及实验过程分析实验结果的合理性。

七、思考题

1. 测定体积吸收系数 $K_Y a$ 及 $\Delta p \sim u$ 有什么实际意义？
2. 如何确定吸收实验的水、氨气、空气的流量？
3. 为什么本实验氨是从气相转移到液相？
4. 理论上本实验过程中气体流量的改变和液体流量的改变对 $K_Y a$ 有何影响？
5. 哪些操作条件会影响到本实验系统的稳定性？
6. 本实验的主要实验误差在哪里？有什么方法可减少误差？

第九章

筛板塔精馏实验

一、实验目的和要求

1. 复习板式精馏塔的基本理论知识，了解板式塔的基本结构、典型流程及附属设备。
2. 掌握板式精馏塔设计方面的知识，掌握精馏过程的基本操作方法。
3. 建立灵敏板的概念，学习判断精馏系统达到稳定的方法。
4. 测定筛板精馏塔全回流和部分回流时的全塔效率。
5. 设计用连续精馏的方法从乙醇溶液中提取乙醇的方案，并在一定时间内完成一定原料量的分离任务。

二、实验原理

板式精馏塔的塔板是气液两相接触的场所，塔釜产生的上升蒸汽与从塔顶下降的液相逐级接触进行传热和传质。下降液经过多次部分汽化，重组分含量逐渐增加，上升蒸汽经多次部分冷凝，轻组分含量逐渐增加，从而使混合物达到一定程度的分离。

1. 全回流时全塔效率（总板效率）E_T 的测定

$$E_T = \frac{N_T - 1}{N_P} \times 100\% \tag{9-1}$$

式中　N_T——完成一定分离任务所需的理论塔板数，包括塔釜；

N_P——完成一定分离任务所需的实际塔板数，本装置 $N_P = 7$。

在全回流操作时，操作线在 $x \sim y$ 图（图 9-1）上为对角线，根据塔顶、塔底的组成在操作线和平衡线间作梯级，即可得到理论塔板数 N_T。

2. 部分回流时全塔效率 E_T' 的测定

（1）精馏段操作线方程

$$y_{n+1} = \frac{R}{R+1} x_n + \frac{x_D}{R+1} \tag{9-2}$$

式中　y_{n+1}——精馏段第 $n+1$ 块塔板上升的蒸汽组成，摩尔分数；

x_n——精馏段第 n 块塔板下降的液体组成，摩尔分数；

R——回流比，$R = L/D$；

x_D——塔顶产品液相组成，摩尔分数。

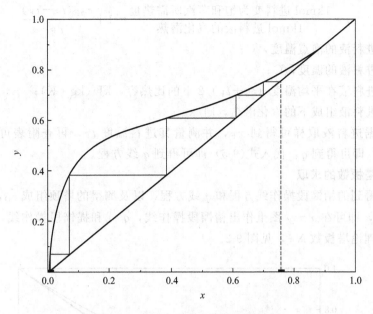

图 9-1 全回流时理论塔板数的确定

本实验中回流量由回流转子流量计测量,但实验操作中一般作冷液回流,故实际回流量需进行校正

$$L = L_0 \left[1 + \frac{c_{pD}(t_D - t_R)}{r_D} \right] \quad (9\text{-}3)$$

式中 L_0——回流转子流量计上的读数值,mL/min;

L——实际回流量,mL/min;

t_D——塔顶液相温度,℃;

t_R——回流液温度,℃;

c_{pD}——塔顶回流液在平均温度 $(t_D+t_R)/2$ 下的比热容,kJ/(kg·K);

r_D——塔顶回流液组成下的汽化潜热,kJ/kg。

产品量 D 可由产品转子流量计测量,由于产品量 D 和回流量 L 的组成和温度均相同,故回流比 R 可直接用两者的比值得到:

$$R = \frac{L}{D} \quad (9\text{-}4)$$

式中 D——产品转子流量计上的读数值,mL/min。

实验中根据塔顶取样分析可得 x_D,测量回流和产品转子流量计读数 L_0 和 D 以及回流温度 t_R 和塔顶液相温度 t_D,再查表可得 c_{pD},r_D,由式(9-3)、式(9-4)可求得回流比 R,代入式(9-2)即可得精馏段操作线方程。

(2) 加料线 (q 线) 方程

$$y = \frac{q}{q-1}x - \frac{x_F}{q-1} \quad (9\text{-}5)$$

式中 q——进料的液相分率;

x_F——进料液的组成,摩尔分数。

$$q = \frac{1\text{kmol 进料变为饱和蒸汽所需热量}}{1\text{kmol 进料液的汽化潜热}} = 1 + \frac{c_{pF}(t_S - t_F)}{r_F} \quad (9\text{-}6)$$

式中　t_S——进料液的泡点温度，℃；

　　　t_F——进料液的温度，℃；

　　　c_{pF}——进料液在平均温度 $(t_S + t_F)/2$ 下的比热容，kJ/(kg·K)；

　　　r_F——进料液组成下的汽化潜热，kJ/kg。

实验中根据进料液取样可得到 x_F，并测量其进料温度 t_F，再查附表可得 t_S、c_{pF} 和 r_F，由式(9-6)即可得到 q，代入式(9-5)即可得到 q 线方程。

(3) 理论塔板数的求取

根据上述得到的精馏段操作线方程和 q 线方程，以及测得的塔顶组成 x_D，塔底组成 x_W 和进料组成 x_F，即可在 $x \sim y$ 图上作出精馏段操作线，q 线和提馏段操作线，然后用 $x \sim y$ 图解法即可得理论塔板数 N_T，见图 9-2。

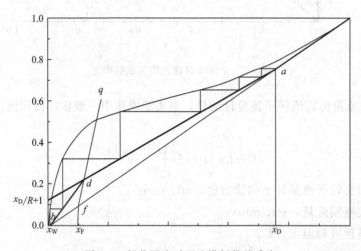

图 9-2　部分回流时理论塔板数的确定

(4) 全塔效率

根据上述求得的理论塔板数 N_T，由式(9-1)便可得到部分回流时的全塔效率 E_T'

$$E_T' = \frac{N_T - 1}{N_P} \times 100\% \quad (9\text{-}7)$$

式中　N_T——完成一定分离任务所需的理论塔板数，包括塔釜；

　　　N_P——完成一定分离任务所需的实际塔板数，本装置 $N_P = 7$。

三、实验装置和流程

精馏塔装置由筛板精馏塔塔釜、塔体（塔板数为7）、冷凝冷却器、加料系统、回流系统、贮槽（原料、产品、釜液）、产品出料管路、残液出料管路、冷却水转子流量计、输液泵以及测量、控制仪表等组成。实验装置流程如图 9-3 所示，虚拟仿真如图 9-4 所示。

筛板精馏塔内径 ϕ68mm，共 7 块塔板，其中精馏段 5 块，提馏段 2 块；精馏段板间距为 150mm，提馏段板间距为 180mm；筛孔孔径 ϕ1.5mm，正三角形排列，空间距 4.5mm，开孔数 104 个。本装置采用电加热方式,塔釜内装有 3 支额定功率为 3kW 的螺旋管加热器。在装置上分别设有料液、产品和釜液的取样口(图中 A、B、C 处)。

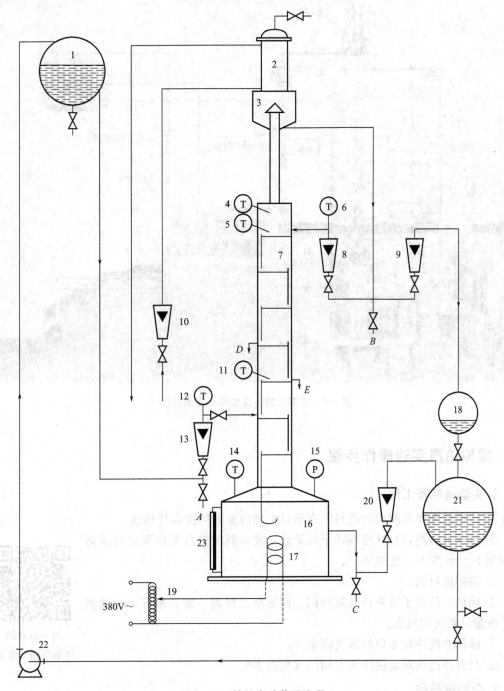

图 9-3 精馏实验装置流程

1—原料贮槽;2—冷凝冷却器;3—集液器;4~6,11,12,14—测温点;7—筛板精馏塔;
8~10,13,20—玻璃转子流量计;15—压力表;16—塔釜;17—加热器;18—产品槽;
19—调压器;21—残液槽;22—输液泵;23—液位计;A~E—采样点

图 9-4 精馏实验装置虚拟仿真示意图

四、虚拟仿真实验操作步骤

1. 实验前准备工作

① 根据浓度要求配成的进料液在原料贮槽内至玻璃液面计顶端。

② 打开进料阀，打开进料转子流量计，使原料液进料至塔釜玻璃液面计顶端下 1cm 左右（见图 9-5）。

③ 关闭进料阀。

④ 确认产品转子流量计及其阀门、回流液采样阀、残液采样阀、残液转子流量计在关闭状态。

⑤ 确认冷凝冷却器顶部排气阀全开。

⑥ 打开冷凝冷却器的冷却水阀门（见图 9-6）。

2. 全回流操作

① 全开回流转子流量计，进行全回流操作（见图 9-7）。

② 开启仪表柜总电源开关。

③ 将电压调节旋钮调节到 190~210V，并保持恒定，等待釜液沸腾（见图 9-8）。

④ 每隔 5min 观察各塔板温度，当灵敏板温度基本不变时，操作即达到稳定。

扫一扫动画
筛板塔精馏实验

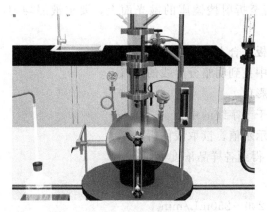

图 9-5 打开进料转子流量计

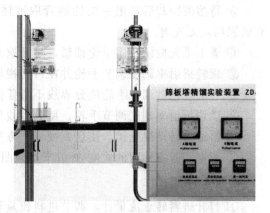

图 9-6 打开冷凝冷却器的冷却水阀门

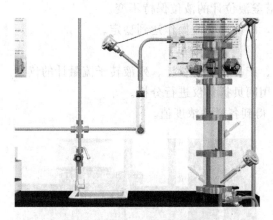

图 9-7 全开回流转子流量计

图 9-8 加热釜液至沸腾

3. 记录数据并分析试样（阿贝折射仪法）

① 记录冷却水转子流量计的读数、回流转子流量计的读数、仪表柜上的数据。
② 同时取馏出液、釜液两个样品（见图 9-9），分析其成分。
③ 将一次性试管放到采样点，打开取样阀，将样品料液加至试管 2/3 处。
④ 将试样试管放入水中冷却，用阿贝折射仪进行分析（见图 9-10）。

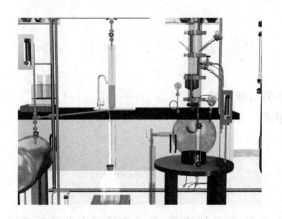

图 9-9 同时取馏出液、釜液两个样品

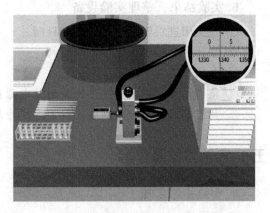

图 9-10 用阿贝折射仪分析样品组成

⑤ 待溶液冷却后，用一次性滴管取液体滴至折射棱镜座的抛光面上，要求液层均匀，充满视场，无气泡。
⑥ 盖上进光棱镜，用手轮锁紧，打开遮光板，合上反射镜。
⑦ 旋转折射率刻度调节手轮并在目镜视场中找到明暗分界线的位置。
⑧ 再旋转色散调节手轮使分界线不带任何彩色。
⑨ 微调折射率刻度调节手轮，使分界线位于十字线的中心。
⑩ 再适当转动聚光镜，使目镜视场下方显示示值，读取液体折射率。
⑪ 根据实验室标定的浓度-折射率对照图，得到各样品浓度值。

4. 部分回流操作

① 打开进料转子流量计，调节进料流量至 200～250mL/min。
② 打开产品阀，调节产品转子流量计使回流比为 3～5（见图 9-11）。
③ 调节残液转子流量计（见图 9-12），使塔釜液位计的液位保持不变。
④ 当釜液液面恒定以及灵敏板温度稳定后，即部分回流操作达到稳定。
⑤ 记录仪表柜上的数据。
⑥ 记录进料转子流量计、回流转子流量计、产品转子流量计、残液转子流量计的读数。
⑦ 分别取进料、馏出液、釜液三个样品，用阿贝折射仪进行分析。
⑧ 根据实验室标定的浓度-折射率对照图，得到各样品浓度值。

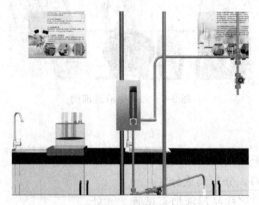

图 9-11　调节产品转子流量计

图 9-12　调节残液转子流量计

5. 实验结束，整理实验设备

① 先关闭进料转子流量计及进料阀。
② 关闭产品转子流量计及产品阀。
③ 关闭残液转子流量计。
④ 将调压器旋钮调至零位，关闭总电源开关切断电源。
⑤ 待乙醇蒸气完全冷凝后，再关冷凝冷却水。

五、实验方法及步骤

1. 实验操作步骤

① 根据浓度要求进行配料（一般 $x_F=0.1$ 左右），并加约 9L 料液于塔釜内至接近玻璃

液面计顶端。若实验前料液已配好，则测定料液组成。

②关闭进、出料阀，关闭采样阀，全开冷凝冷却器顶部排气阀。稍开冷凝冷却水阀门，全开回流转子流量计阀门，进行全回流操作。

③开启仪表柜总电源开关，将电压调节旋钮调节到所需要的加热电压并保持恒定。

④待釜液开始沸腾，开大冷凝冷却水阀门到转子流量计读数最大值，并保持恒定。

⑤加热电压和冷凝冷却水量都维持恒定后，每隔5min观察各塔板温度，当灵敏板温度计11的读数基本不变时，操作即达到稳定。分别取馏出液、釜液两个样品，分析组成，读取回流液和冷凝冷却水的流量，并分别读取精馏塔中的六个温度。

⑥部分回流操作。首先将进料转子流量计调到200~250mL/min，其次通过对产品转子流量计的调整，将回流比调至3~5，最后通过对釜液转子流量计的调整，使塔釜液位计的液位保持不变。当釜液液面恒定以及灵敏板温度稳定后，即部分回流操作达到稳定。读取各转子流量计的流量和各温度计的温度，并测取馏出液、料液和釜液样品的组成。

⑦实验结束。先关闭进料液、馏出液、釜液的流量调节阀，再将调压器旋钮调至零位，关闭总电源开关切断电源，待乙醇蒸气完全冷凝后，再关冷凝冷却水，并做好清洁工作。

2. 注意事项

①实验开始时，应先向塔顶冷凝冷却器通冷却水，后给塔釜加热；结束时反之。

②塔釜内料液的位置应始终高于加热釜电极的高度。

六、实验报告

1. 将塔顶、塔底温度和组成，以及各流量计读数等原始数据列表。
2. 按全回流和部分回流分别用图解法计算理论板数。
3. 计算全塔效率。
4. 分析并讨论实验过程中观察到的现象。

七、思考题

1. 影响精馏操作稳定的因素是哪些？维持塔稳定操作应注意哪些？如何判断塔的操作已达到稳定？
2. 在全回流条件下，改变加热功率对塔的分离效果有何影响？
3. 塔顶冷回流对塔内回流液量有何影响？如何校正？
4. 用转子流量计来测定乙醇水溶液流量，计算时应怎样校正？
5. 影响全塔效率的主要因素有哪些？

第十章

萃取实验——转盘萃取、脉冲萃取

一、实验目的和要求

1. 了解转盘萃取塔和脉冲萃取塔的基本结构、操作方法及萃取的工艺流程。
2. 观察转盘萃取塔转盘转速变化时或脉冲萃取塔脉冲参数变化时，萃取塔内轻、重两相流动状况，了解萃取操作的主要影响因素，研究萃取操作条件对萃取过程的影响。
3. 测量传质单元数、传质单元高度和体积传质系数K_{YV}，关联传质单元高度与脉冲萃取过程操作变量的关系。
4. 计算萃取率 η。

二、实验原理

萃取是分离和提纯物质的重要单元操作之一，是利用混合物中各个组分在外加溶剂中的溶解度的差异而实现组分分离的单元操作。进行液-液萃取操作时，两种液体在塔内作逆流流动，其中一液体作为分散相，以液滴的形式通过另一作为连续相液体，两种液相浓度在设备内作微分式的连续变化，并依靠密度差在塔的两端实现两液相间的分离。当轻相作为分散相时，相界面出现在塔的上部；反之相界面出现在塔的下端。本实验以轻相为分散相，相界面出现在塔的上部。

计算微分逆流萃取塔的塔高时，主要是采取传质单元法。即以传质单元数和传质单元高度来表征，传质单元数表示过程分离程度的难易，传质单元高度表示设备传质性能的好坏。

1. 萃取的基本符号（见表 10-1）

表 10-1 萃取的基本符号

名称	符号	流量单位	组成符号
原料液	F	kg/s	X_F 或 x_F
萃余相	R	kg/s	X_R 或 x_R
萃取剂	S	kg/s	Y_S 或 y_S
萃取相	E	kg/s	Y_E 或 y_E

2. 萃取的物料衡算

如图 10-1、图 10-2 所示,萃取计算中各相组成可用操作线方程相关联,操作线上的点 $P(X_R, Y_S)$ 和点 $Q(X_F, Y_E)$ 分别与装置的上部和下部相对应。

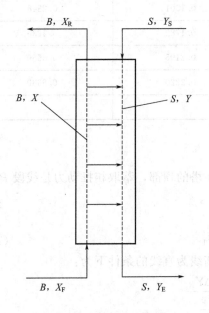

图 10-1 物料衡算示意图

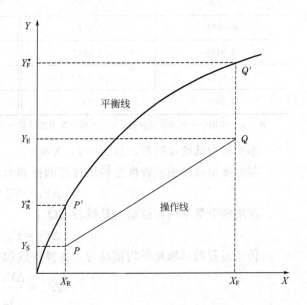

图 10-2 平均推动力计算示意图

在第一溶剂 B 与萃取剂 S 完全不互溶时,萃取过程的操作线在 $X \sim Y$ 坐标系中为直线,其方程式见式(10-1):

$$\frac{Y_E - Y_S}{X_F - X_R} = \frac{Y - Y_S}{X - X_R} \tag{10-1}$$

化简得

$$Y - Y_S = m(X - X_R)$$

其中

$$m = \frac{Y_E - Y_S}{X_F - X_R}$$

单位时间内从第一溶剂中萃取出的纯物质 A 的量 M,可由物料衡算确定:

$$M = B(X_F - X_R) = S(Y_E - Y_S) \tag{10-2}$$

3. 萃取过程的质量传递

不平衡的萃取相与萃余相在塔的任一截面上接触,两相之间发生质量传递。物质 A 以扩散的方式由萃余相进入萃取相,该过程的极限是达到相间平衡,相平衡的相间关系为:

$$Y^* = kX \tag{10-3}$$

k 为分配系数,只有在较简单的体系中,k 才是常数,一般情况下为变数。本实验采用煤油-苯甲酸-水体系,给出如表 10-2 所示的系统平衡数据。

表 10-2 煤油-苯甲酸-水系统在室温下的平衡数据表

y（质量分数）	x（质量分数）	y（质量分数）	x（质量分数）
0.00972	0.0128	0.1144	0.1786
0.0195	0.0199	0.1301	0.2348
0.0354	0.0270	0.1782	0.4230
0.0683	0.0817	0.2195	0.6550
0.0725	0.0990	0.2220	0.6330
0.1010	0.1494		

注：x—油相中苯甲酸质量分数；y—水相中苯甲酸质量分数。

本实验为低浓度过程，故 $Y \approx y$，$X \approx x$。

与平衡组成的偏差程度是传质过程的推动力，在塔的顶部，萃取相推动力是线段 PP'：

$$\Delta Y_R = Y_R^* - Y_S \tag{10-4}$$

在塔的下部萃取相推动力是线段 QQ'：

$$\Delta Y_F = Y_F^* - Y_E \tag{10-5}$$

传质过程的萃取相平均推动力，在操作线和平衡线为直线的条件下为：

$$\Delta Y_m = \frac{\Delta Y_F - \Delta Y_R}{\ln \frac{\Delta Y_F}{\Delta Y_R}} \tag{10-6}$$

物质 A 由萃余相进入萃取相的过程的传质动力学方程式为：

$$M = K_Y A \Delta Y_m \tag{10-7}$$

式中　K_Y——萃取相传质系数，kg/(m²·s)；
　　　A——相接触表面积，m²。

该方程式中的萃取塔内相接触表面积 A 不能确定，因此通常采用另一种方式。
相接触表面积 A 可以表示为：

$$A = aV = a\Omega h \tag{10-8}$$

式中　a——相接触比表面积，m²/m³；
　　　V——萃取塔有效操作段体积，m³；
　　　Ω——萃取塔横截面积，m²；
　　　h——萃取塔操作部分高度，m。

这时

$$M = K_Y a V \Delta Y_m = K_{YV} V \Delta Y_m \tag{10-9}$$

式中　$K_Y a = K_{YV}$——体积传质系数，kg/(m³·s)。

根据式(10-2)、式(10-7)、式(10-8) 和式(10-9)，可得

$$h = \frac{S}{K_{YV}\Omega} \times \frac{Y_E - Y_S}{\Delta Y_m} = H_{OE} N_{OE} \tag{10-10}$$

式中　$H_{OE} = \frac{S}{K_{YV}\Omega}$，称为萃取相传质单元高度；

　　　$N_{OE} = \frac{Y_E - Y_S}{\Delta Y_m}$，称为萃取相传质单元数。

K_Y、K_{YV}、H_{OE} 是表征质量交换过程特性的，K_Y、K_{YV} 越大，H_{OE} 越小，则萃取过程进行得越快。

$$K_{YV}=\frac{M}{V\Delta Y_m}=\frac{S(Y_E-Y_S)}{V\Delta Y_m} \tag{10-11}$$

4. 萃取率

$$\eta=\frac{被萃取剂萃取的组分\,A\,的量}{原料液中组分\,A\,的量}\times 100\%$$

所以

$$\eta=\frac{S(Y_E-Y_S)}{BX_F}\times 100\% \tag{10-12}$$

或

$$\eta=\frac{B(X_F-X_R)}{BX_F}\times 100\%=\left(1-\frac{X_R}{X_F}\right)\times 100\% \tag{10-13}$$

5. 质量流量和组成

① 第一溶剂 B 的质量流量

$$B=F(1-x_F)=V_F\rho_F(1-x_F) \tag{10-14}$$

式中　B——溶剂的质量流量，kg/h；
　　　F——料液的质量流量，kg/h；
　　　V_F——料液的体积流量，m³/h；
　　　ρ_F——料液的密度，kg/m³；
　　　x_F——料液中 A 的含量，kg/kg。

液体流量计校正：V_F 由式(10-15) 计算：

$$V_F=V_N\sqrt{\frac{\rho_0(\rho_f-\rho_F)}{\rho_F(\rho_f-\rho_0)}}\approx V_N\sqrt{\frac{\rho_0}{\rho_F}} \tag{10-15}$$

式中　V_N——转子流量计读数，mL/min 或 m³/h；
　　　ρ_f——转子密度，kg/m³；
　　　ρ_0——20℃时水的密度，kg/m³。

所以

$$B=V_N\sqrt{\rho_0\rho_F}(1-x_F) \tag{10-16}$$

② 萃取剂（水）S 的质量流量

$$S=V_S\rho_S \tag{10-17}$$

式中　V_S——萃取剂（水）体积流量，m³/h；
　　　ρ_S——萃取剂（水）的密度，kg/m³。

③ 原料液及萃余相的组成 x_F、x_R

对于煤油-苯甲酸-水体系，采用酸碱中和滴定的方法可测定原料液组成 x_F、萃余相组成 x_R 和萃取相组成 y_E，即苯甲酸的质量分率，y_E 也可通过如上的物料衡算而得，具体步骤如下：

用移液管取试样 V_1 mL，加指示剂 1～2 滴，用浓度为 N_b 的 NaOH 溶液滴定至终点，如用去 NaOH 溶液 V_2 mL，则试样中苯甲酸的摩尔浓度 N_a 为：

$$N_a = \frac{V_2 N_b}{V_1} \tag{10-18}$$

则

$$x_F = \frac{N_a M_A}{\rho_F} \tag{10-19}$$

式中　M_A——溶质 A 的分子量，g/mol，本实验中苯甲酸的分子量为 122g/mol。
　　　ρ_F——溶液密度，g/L。

x_R 亦用同样的方法测定：

$$x_R = \frac{N_a' M_A}{\rho_R} \tag{10-20}$$

其中

$$N_a' = \frac{V_2' N_b}{V_1'} \tag{10-21}$$

式中　V_1'，V_2'——样品的体积与滴定所耗的 NaOH 溶液的体积。

三、实验装置和流程

1. 转盘萃取塔

主要设备是转盘萃取塔，塔体是内径为 50mm 的玻璃管，塔顶电机连接转轴，转轴上固定有圆盘，塔壁上固定有圆环，圆环与圆盘交错布置，转盘萃取实验流程及虚拟仿真见图 10-3、图 10-4。

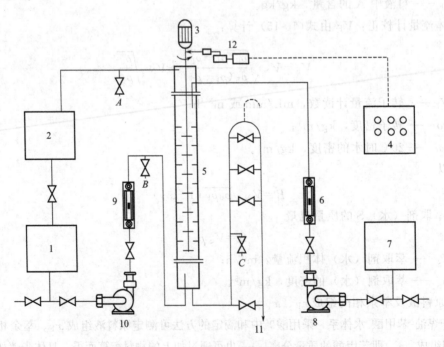

图 10-3　转盘萃取实验流程

1—原料贮槽（苯甲酸-煤油）；2—收集槽（萃余液）；3—电机；4—控制柜；5—转盘萃取塔；6,9—转子流量计；7—萃取剂贮罐（水）；8,10—输送泵；11—排出液（萃取液）管；12—转速测定仪；$A \sim C$—取样口

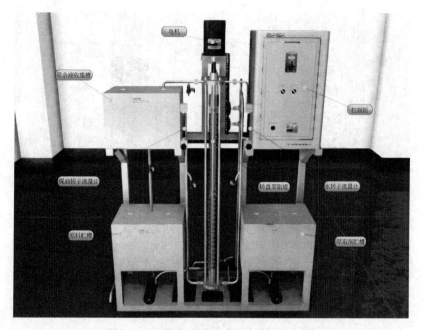

图 10-4　转盘萃取实验装置虚拟仿真示意图

2. 脉冲萃取塔

主要设备是脉冲萃取塔，塔体是内径为 50mm 的玻璃管，内装不锈钢丝网填料，脉冲萃取实验流程及虚拟仿真见图 10-5、图 10-6。

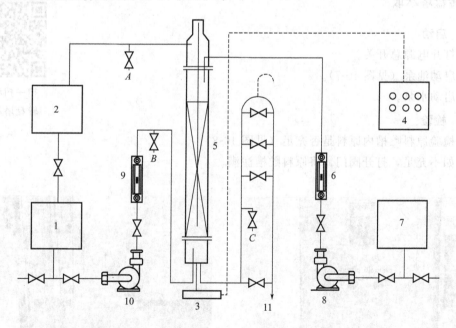

图 10-5　脉冲萃取实验流程

1—原料贮槽（苯甲酸-煤油）；2—收集槽（萃余液）；3—脉冲系统；4—控制柜；
5—填料（脉冲）萃取塔；6,9—转子流量计；7—萃取剂贮罐（水）；
8,10—输送泵；11—排出液（萃取液）管；$A \sim C$—取样口

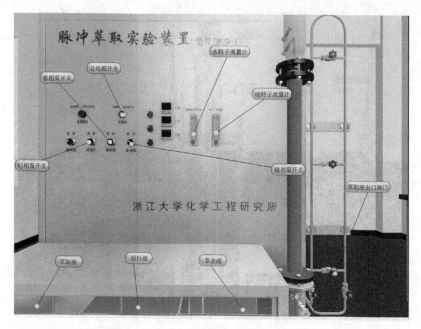

图 10-6 脉冲萃取实装置虚拟仿真示意图

四、虚拟仿真实验操作步骤

1. 转盘塔萃取

(1) 启动

① 打开电源总开关。

② 启动油泵（见图 10-7）。

③ 启动水泵。

(2) 检验

① 检验原料贮槽内原料是否充足（见图 10-8）。

② 如不充足，打开阀门，将原料贮槽注满。

扫一扫动画
转盘塔萃取

图 10-7 控制柜操作

图 10-8 检验原料贮槽内原料是否充足

③ 检验萃取剂贮槽内的水是否充足。

④ 如不充足，打开水阀门，将萃取剂贮槽注满。

(3) 当转速为0的时候，进行实验

① 调节重相（水）转子流量计，流量在 100～200mL/min（见图 10-9）。

② 水在萃取塔内流动 5min 后，调节轻相——煤油转子流量计，流量在 100～200mL/min。

③ 调节萃取相出口阀门，直至保持安静区两相分界面的恒定（见图 10-10）。

④ 实验稳定约 30min 后，打开取样阀进行取样分析。

图 10-9　调节重相——水流量计

图 10-10　油、水相分界面

(4) 分析原料液

① 移出试管，打开原料液取样阀门，取液体至试管的 2/3 处（见图 10-11）。

② 取出密度计，按照量程从小到大选择密度计开始测量（见图 10-12），记录密度计的读数。

图 10-11　取原料液样品

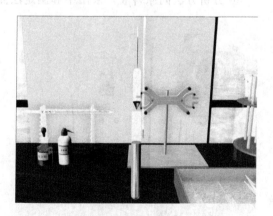
图 10-12　测样品的密度

③ 用移液管取 10mL 原料液至锥形瓶（见图 10-13）。

④ 滴 2 滴指示剂在锥形瓶中，充分振荡。

⑤ 碱式滴液管内放入氢氧化钠溶液。

⑥ 滴加氢氧化钠溶液的同时充分振荡锥形瓶（见图 10-14），直至锥形瓶中的液体颜色由黄变绿为止。

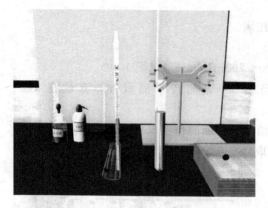
图 10-13 用移液管移取液体

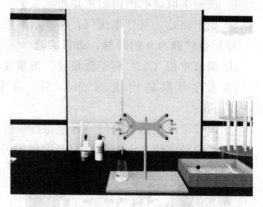
图 10-14 滴定分析

⑦ 若液体颜色变为蓝色，则说明氢氧化钠过量，需要重新取样分析。
⑧ 记录碱式滴液管滴定用量。

(5) 分析萃余相
① 移出试管，打开萃余相取样阀门，取萃余相样品至试管的 2/3 处。
② 萃余相的分析方法同原料液，即采用中和滴定法测定。

(6) 改变转速，重复实验
① 打开转速开关（见图 10-15），调节转速。
② 水和煤油流量保持不变，观察塔顶安静区相分界面的位置。
③ 调节萃取相出口阀门，直至保持安静区两相分界面的恒定（见图 10-16）。
④ 实验稳定约为 30min。
⑤ 两相分界面恒定的情况下，打开萃余相取样阀进行取样分析。
⑥ 分析方法同原料液，采用中和滴定法测定。

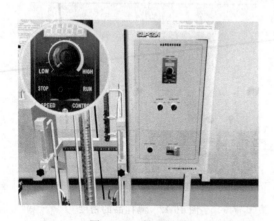
图 10-15 转速调节
图 10-16 调节萃取相出口阀门 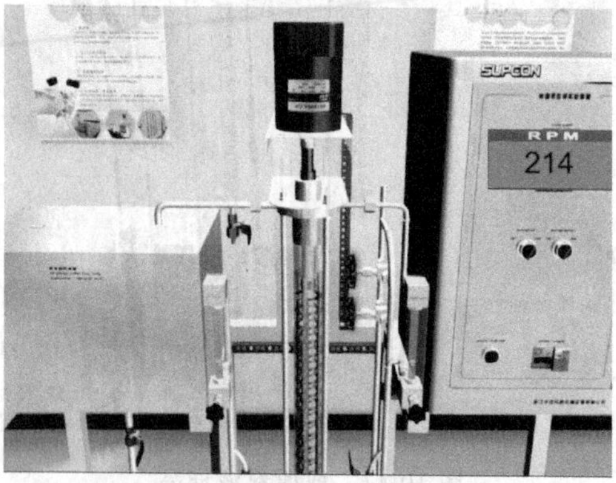

(7) 关闭设备，清洗仪器
① 关闭转速调速阀、转速开关。
② 关闭水转子流量计、煤油转子流量计。
③ 关闭油泵开关、水泵开关及总开关。

2. 脉冲塔萃取

(1) 检查装置，开始实验

① 实验前先检查萃取剂与原料液两贮槽内溶液是否充足，若不足则补充。

② 打开总电源、重相泵、轻相泵开关（见图10-17）。

③ 打开萃取剂转子流量计，向塔内加入萃取剂，充满全塔。

④ 打开原料液转子流量计，根据萃取剂比例，加入原料液（见图10-18）。

扫一扫动画
脉冲塔萃取

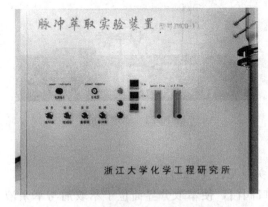

图10-17 脉冲萃取操作控制柜

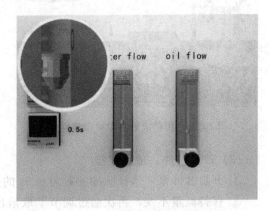

图10-18 打开原料液转子流量计

(2) 不开启脉冲时，取样分析

① 调节萃取液出口阀门，使流量保持不变，油、水相分界面位于萃取剂与萃余液出口之间。

② 稳定30min，打开萃余液取样阀，取样至试管2/3处（见图10-19）。

③ 选用合适的密度计测定萃余液，读取密度计读数（见图10-20）。

图10-19 萃余液取样

图10-20 读取密度计读数

④ 用中和滴定法测定萃余液的组成。

⑤ 取足量NaOH溶液至滴定管中，用移液管取10mL萃余液至锥形瓶中（见图10-21），向锥形瓶中加入指示剂。

第十章 萃取实验——转盘萃取、脉冲萃取

⑥ 滴定时应注意缓慢滴定，并充分振荡，直至溶液变成绿色（见图 10-22）。
⑦ 另取一根新试管，打开原料液取样阀，取样至试管 2/3 处。
⑧ 原料液测定方法和步骤与萃余液相同。
⑨ 测定结束后将试管内剩余原料液倒回原料贮槽内。

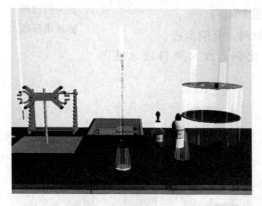

图 10-21　用移液管移取萃余液样品

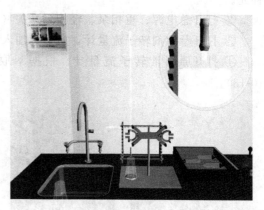

图 10-22　滴定分析

(3) 改变脉冲，重复实验

① 开启脉冲泵，选择脉冲参数为 0.5s 的开关（见图 10-23）。
② 保持流量不变，再次通过调节萃取液出口阀门，使萃取分界面位于萃取剂与萃余液出口之间（见图 10-24）。
③ 稳定 30min，测量并记录脉冲参数，测定萃余液的密度及组成。
④ 改变脉冲参数重复实验。

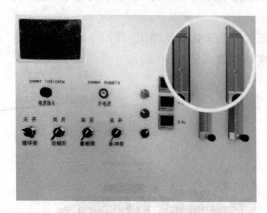

图 10-23　打开脉冲泵开关

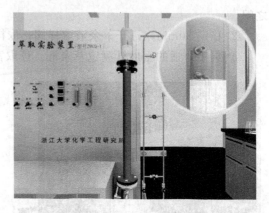

图 10-24　油、水相分界面

(4) 实验结束后，整理实验装置

① 关闭脉冲系统，关闭脉冲泵。
② 关闭重相泵，关闭轻相泵。
③ 关闭两个转子流量计。
④ 关闭萃取液出口阀门。
⑤ 最后关闭总电源开关。

五、实验方法及步骤

1. 实验操作步骤

(1) 转盘萃取塔

① 配制原料液：煤油-苯甲酸溶液，浓度约为 0.3%（质量分数），体积约 50L，加入原料贮槽中。

② 将萃取剂（蒸馏水）加入萃取剂贮罐中。

③ 打开流量调节阀，水在萃取塔内流动 5min 后，开启分散相——油相管路，调节两相流量在 100~200mL/min，待分散相在塔顶凝聚一定厚度后，再通过调节连续相出口阀，以保持安静区中两相分界面的恒定。

④ 通过调节转速来控制外加能量的大小，在操作中逐渐增大转速，一般取 100~700r/min。

⑤ 每次实验稳定时间约 30min，然后打开取样阀取样，用密度计和中和滴定法测定萃余液及原料液的密度和组成，同时记录转速。

⑥ 改变转速，重复上述实验。

⑦ 实验结束后，将实验装置恢复原样。

(2) 脉冲萃取塔

① 配制原料液：煤油-苯甲酸溶液，浓度约为 0.3%（质量分数），体积约 50L，加入原料贮槽中。

② 将萃取剂（水）加入萃取剂贮罐中。

③ 打开流量调节阀，先向塔内加入萃取剂，充满全塔，并至所需流量。加入原料液，根据萃取剂比例调节流量。在实验过程中保持流量不变，并通过调节萃取液出口阀门，以保持安静区中两相分界面的恒定（见图 10-10）。

④ 打开脉冲系统，选择所需脉冲频率。

⑤ 每次实验稳定时间约 30min，然后打开取样阀取样，用密度计和中和滴定法测定萃余液及原料液的密度和组成，同时记录脉冲参数。

⑥ 改变脉冲参数，重复上述实验。

⑦ 实验结束后，将实验装置恢复原样。

2. 注意事项

① 调节转速时要小心谨慎，缓慢升速，不可增速过快。从流体力学性能考虑，若转速太高，容易液泛，操作不稳定。

② 在整个实验过程中，维持塔顶两相分界面稳定在轻相出口和重相入口之间适中的位置。

③ 由于分散相和连续相在塔内存在滞留，改变操作条件后，需要足够长稳定时间，否则误差极大。

④ 煤油的实际体积流量并不等于流量计的读数。煤油的实际流量必须用流量校正式对流量计的读数进行修正。

六、实验报告

1. 计算不同转速/脉冲下的传质单元数 N_{OE}；
2. 计算不同转速/脉冲下的传质单元高度 H_{OE}。
3. 计算不同转速/脉冲下的体积传质系数 $K_Y a$。
4. 根据传质理论对上述实验结果做出分析。

七、思考题

1. 分析比较萃取实验装置与吸收、精馏实验装置的异同点？
2. 采用本实验的萃取装置如何调节外加能量和进行测量？
3. 从实验结果分析转盘转速变化或脉冲参数变化对萃取传质系数与萃取率的影响。
4. 测定原料液、萃取相、萃余相组成可用哪些方法？
5. 萃取过程是否会发生液泛，如何判断液泛？

第十一章 洞道式干燥特性曲线测定实验

一、实验目的和要求

1. 了解洞道式干燥装置的结构、流程及其操作方法。
2. 作出物料在恒定干燥条件下的干燥特性曲线（$X\sim\tau$，$U\sim X$），并求出临界含水量 X_c、平衡含水量 X^* 及恒速阶段的干燥速率 $U_{恒速}$。
3. 求出恒速阶段的传质系数 k_W 和传热系数 α。
4. 改变气温或气速等操作条件，作出不同空气参数下的干燥特性曲线，同时求出各自的临界含水量、平衡含水量以及恒速阶段的干燥速率、传质系数和传热系数。

二、实验原理

物料中所含湿分性质不同，反映在物料的干燥上，其过程的变化也必各异。为了减少影响因素，我们将湿物料在恒定干燥条件下（即干燥介质空气的温度、湿度、速率以及与物料接触的方式均维持恒定）进行干燥实验，实验中，通过对湿物料在不同时间内重量的称量，即可求得干燥时间 τ 与湿物料重量 G 的关系，将数据加以整理可得物料的干燥曲线 $X\sim\tau$ 和干燥速率曲线 $U\sim X$。

考察干燥曲线和干燥速率曲线可知，整个干燥过程主要有恒速干燥与降速干燥两个阶段。

恒速干燥阶段：湿物料表面全部为非结合水所润湿。在物料表面水分汽化过程中，湿物料内部水分向表面扩散的速率等于或大于水分的表面汽化速率，物料表面总是维持湿润状态，且其表面温度亦为该空气状态下的湿球温度 t_W，故该阶段又称为表面汽化控制阶段。

降速干燥阶段：湿物料内部水分向表面扩散的速率低于物料表面的汽化速率，则物料温度升高或表面变干，进入降速阶段。随着物料不断干燥，其内部水分愈来愈少，这样，水分由内部向表面传递的速率愈来愈慢，干燥速率亦随之降低，直至物料含水量达到该空气状态下的物料平衡含水量 X^*。由于降速阶段的干燥速率取决于物料本身的结构、形状和尺寸，而与干燥介质状况关系不大，故降速阶段又称物料内部迁移控制阶段。临界含水量 X_c，即为恒速与降速阶段的转折点，临界含水量对于干燥过程研究和干燥器的设计都是十分重要的。

1. 干燥特性曲线 ($X \sim \tau$，$U \sim X$) 的求取

干燥速率定义为单位时间、单位干燥表面积所除去的湿分重量，用 U 表示，故：

$$U = \frac{dW}{A d\tau} = -\frac{G_c}{A} \times \frac{dX}{d\tau} \tag{11-1}$$

式中　U——干燥速率，$kg/(m^2 \cdot s)$；
　　　A——干燥面积，m^2；
　　　G_c——物料绝干重量，kg；
　　　X——物料干基含水量，kg 水$/kg$ 干空气。

根据计算机和重量传感器测出的不同时刻物料重量与时间的关系曲线 $G_i \sim \tau$，可得出不同时刻物料的干基含水量 X_i：

$$X_i = \frac{G_i - G_c}{G_c} \tag{11-2}$$

式中　X_i——物料在 τ_i 时刻含水量，kg 水$/kg$ 干料；
　　　G_i——τ_i 时刻物料量（包括附件重），kg；
　　　G_c——物料绝干重量（包括附件重），kg。

按式(11-2) 可得 τ_i 时刻所对应的 X_i 值，据此即可作出干燥曲线 $X \sim \tau$，从 $X \sim \tau$ 曲线图中可找出 X_c，再在 $X \sim \tau$ 曲线上取代表性的点作图求出斜率 $\frac{dX}{d\tau}$，再按式(11-1) 即可计算出干燥速率U_i，然后绘出干燥速率曲线 $U \sim X$，从 $U \sim X$ 图中可以找出 X^* 和 $U_{恒速}$。

2. 恒定干燥条件下的传质系数与传热系数的测定

若单位时间内从单位面积物料表面传递到空气主体的水分量为 G_W，则

$$G_W = k_W(H_W - H) \tag{11-3}$$

根据传质速率方程，在恒速干燥阶段：

$$U_{恒速} = G_W = k_W(H_W - H) \tag{11-4}$$

故：

$$k_W = \frac{U_{恒速}}{H_W - H} \tag{11-5}$$

若单位时间内从空气主体传递到单位面积物料表面的热量为 Q，则

$$Q = \alpha(t - t_W) \tag{11-6}$$

根据"汽化所需热量等于空气与湿物料间的对流传热量"进行热量衡算，即

$$Q = r_W U_{恒速} = r_W k_W(H_W - H) \tag{11-7}$$

所以

$$\alpha(t - t_W) = r_W U_{恒速} = r_W k_W(H_W - H) \tag{11-8}$$

即

$$\alpha = \frac{r_W U_{恒速}}{t - t_W} = \frac{r_W k_W(H_W - H)}{t - t_W} \tag{11-9}$$

式中　$U_{恒速}$——恒速干燥阶段的干燥速率，$kg/(m^2 \cdot s)$；
　　　Q——单位时间内从空气主体传递到单位面积物料表面的热量，W/m^2；
　　　G_W——单位时间内从单位面积物料表面传递到空气主体的水分量，$kg/(m^2 \cdot s)$；
　　　k_W——恒速干燥阶段的传质系数，$kg/(m^2 \cdot s)$。

H —— 空气湿度，kg 水/kg 干空气；

H_W —— 湿球温度 t_W 下空气的饱和湿度，kg 水/kg 干空气；

α —— 恒速干燥阶段物料表面与空气之间的对流传热系数，W/(m²·℃)；

t_W —— 干燥器内空气的湿球温度，℃；

t —— 干燥器内空气的干球温度，℃；

r_W —— t_W 下水的汽化潜热，J/kg。

在本实验中，测出物料失重与时间的关系 $G_i \sim \tau$，即可得出 $X \sim \tau$，$U \sim X$ 曲线，进而求出临界含水量 X_c、平衡含水量 X^* 和 $U_{恒速}$，并最终求出恒速干燥阶段的传质系数 k_W 和传热系数 α。

三、实验装置和流程

实验装置如图 11-1 所示，风机将空气送入预热室进行预热，冷空气经电加热到温度 T_1 后，进入干燥室将热能供给物料，然后直接排放至大气。

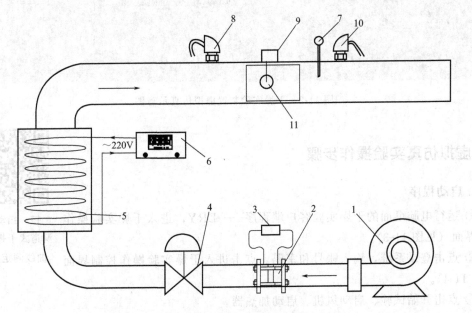

图 11-1 干燥实验装置流程

1—风机；2—孔板流量计；3—压差变送器；4—电动调节阀；5—加热器；6—温控系统；
7—湿球温度计；8,10—干球温度计；9—重量传感器；11—干燥物料

空气的流量由孔板流量计测量，孔板两端压差用压差变送器测量，空气流量由电动调节阀经计算机在线控制调节。系统内空气温度由铜-康铜热电偶温度计测定，干燥室内空气入口及出口的干球温度由热电偶温度计 8、10 测量，湿球温度计 7 测量干燥室出口的湿球温度。空气进口温度 T_1 采用计算机自动控制。物料重量变化由重量传感器 9 测量并由计算机检测显示。

干燥实验装置虚拟仿真如图 11-2 所示。

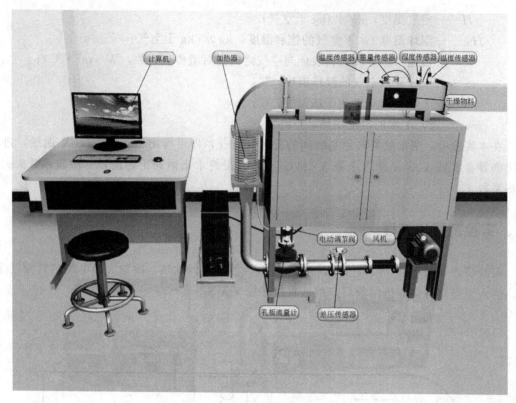

图 11-2 干燥实验装置虚拟仿真示意图

四、虚拟仿真实验操作步骤

1. 启动程序

① 运行电脑桌面的干燥实验客户端程序——DRY，进入干燥实验操作登录界面（见图 11-3）。

② 点击登录系统，输入账号和密码，点击进入干燥实验操作控制界面（见图 11-4）。

扫一扫动画
洞道式干燥特性
曲线测定实验

③ 点击开始试验、启动风机、启动加热器。

④ 该实验以改变空气的流量来进行，建议空气流量为 0.025kg/s 和 0.06kg/s，温度保持在 50℃。

⑤ 设置干燥室温度为 50℃，空气流量为 0.025kg/s，建议采样时间为 60s/次。

⑥ 待空气温度显示为一条直线且跳出加水提示后，可对干燥物进行加水浸湿。

2. 加水浸湿

① 使用量杯对湿球温度计下面的烧杯进行加水，直到将水加满为止（见图 11-5）。

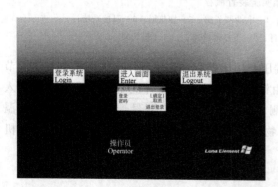

图 11-3 干燥实验操作登录界面

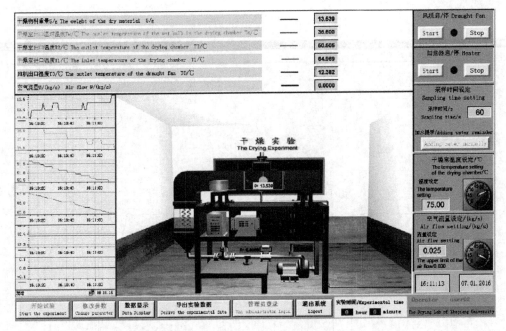

图 11-4　干燥实验操作控制界面

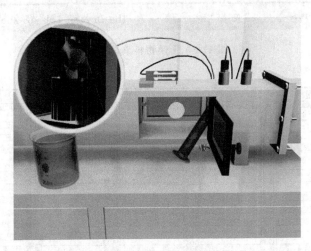

图 11-5　对湿球温度计进行加水

② 取出干燥物料，浸泡在烧杯里 2～3min，将物料放回。
③ 物料的重量不再明显减轻时，改变实验条件继续实验（见图 11-6）。

3. 改变参数，重复实验

① 将空气流量增加到 0.06kg/s，继续实验。
② 继续观察实验，直至物料重量不再明显减轻。
③ 保存实验数据。

4. 保存数据

① 点击屏幕"导出实验数据"（见图 11-7），选择屏幕左上角的导出图标。
② 选择全部实验，单击确定按钮，导出数据，数据保存在桌面。

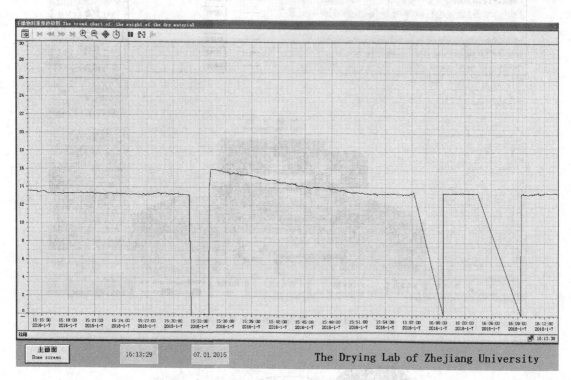

图 11-6 干燥物料重量趋势

图 11-7 数据记录

5. 关闭程序

① 点击主屏幕，回到运行界面。
② 关闭加热器。
③ 待干球温度降至常温后关闭风机电源。
④ 点击退出系统。

五、实验方法及步骤

1. 实验操作步骤

① 打开计算机，运行干燥实验程序，登录后进入干燥实验操作控制界面。
② 点击开始实验按钮，先启动风机，将风量设定到最大值，再开加热器，将空气进口温度设定在某一合适值（一般设置为 60～90℃），待空气进口温度基本不变时，可对干燥物料进行加水浸湿，同时对湿球温度计下面的水杯进行加水，加水结束后，干燥即可自动开始，当物料的重量不再明显减轻时，表示该干燥条件下的干燥已经结束。
③ 改变空气流量或温度，重复上述实验。
④ 实验测试完毕后，将实验数据导出保存。
⑤ 实验结束。先关闭加热电源，开大风量，待空气进口温度降至 50℃ 以下后，再关闭风机，退出计算机实验操作控制界面。

2. 注意事项

称重传感器易损坏，使用时需小心，不能超重。

六、实验报告

1. 绘制干燥曲线和干燥速率曲线图。
2. 求出物料的临界含水量。
3. 计算恒速干燥阶段时物料表面与空气之间对流传热系数。

七、思考题

1. 本实验如何做到恒定干燥条件？
2. 实验中空气的湿度 H、H_W 如何求得？
3. 改变气流速率或温度，临界含水量、平衡含水量如何改变？
4. 空气循环式干燥装置和废气排放式干燥装置各有什么优缺点？

Chemical Engineering Principle Experiments and Virtual Simulation (Bilingual)

Xiangqun Ye   Yan Shan   Editor-in-chief

Abstract

This textbook is prepared based on the teaching requirements of chemical engineering principle for chemical engineering students and relevant professions in comprehensive universities and previous lectures on experiment teaching for chemical engineering principle. The content includes introduction, methods of experimental study on chemical engineering principle, processing technique of experimental data as well as an overview for virtual simulation of experiments on chemical engineering principle. The details of the experiments include: experiment principle, experimental set up, operating procedure and virtual simulation for the determination of comprehensive fluid mechanics resistance, centrifugal pump characteristic curve, convective heat transfer coefficient, filtration at constant pressure and vacuum, absorption in packed column, rectification in sieve-plate column, rotating disc and pulsed extraction, and determination of the characteristic curve in tunnel dryer.

With the rapid development of innovative educational technology, virtual simulation has been an effective approach for experimental teaching. Simulated experiment environments and objects with high emulation were established for experimental teaching and they rely on the technologies of virtual reality, multimedia, human-computer interaction, database and network communication. In this book, supporting materials such as new three-dimensional teaching resources and tools with various forms were provided to meet the requirements of innovative teaching methodologies and improve the efficiency of instruction significantly.

This book can be used as a teaching material for experimental module of chemical engineering principle and basic chemical engineering experiment for the students in undergraduate colleges and also be used as a reference for technical staff in chemical engineering, chemistry, etc.

Preface

Chemical engineering principle experiment is an important fundamental course for students in chemical engineering and relevant majors. Moreover, it is often regarded as the necessity of bridging the gap between academia and industry as well as an important way for training students' engineering technology capabilities.

The teaching method in conventional chemical engineering was adopted from theoretical knowledge, operation experiences and practices. However, many drawbacks make it difficult to teach experiments such as high cost of experimental devices, operation complexity, space, high pollution, risks, etc.

With the rapid development of science and technology, chemical engineering experimental teaching moved to highly simulated virtual experimental environments that use virtual technologies, human-computer interaction, and network communication. Nevertheless, the introduction of virtual simulation into chemical engineering experiments has significantly improved the teaching efficiency and effectiveness. Therefore, the advantages of virtual simulation can be illustrated as follows.

① Virtual simulation experiment software can help in achieving educational outcomes and enhance students' learning interest by shifting from actual operation to simulation training (virtual space) in a computer. This software enables students to have a solid background and significantly reduces experiment operational risks before taking class.

② Virtual simulation experiment has no limitations of time and space and experimental teaching can be carried out anytime and anywhere;

③ Virtual simulation can provide practical teaching guide for teachers and learning institutions in teaching students.

This book is a teaching guide, and useful material for chemical engineering experiments and it is emphasizing on the combination of theory, training, and practice of experimental quality. Also, it guides and enables students on how to operate the actual experimental devices independently after performing virtual simulation experiments, designing experiments, collecting, and analyzing experimental data, etc.

The chief editors of this book are Ye Xiangqun and Shan Yan, while other contributors are Yang Guocheng, Jin Weiguang, Dou Mei and Nan Suifei. The checking work is conducted by Wan Wenjing and Niu Caoping. Due to limited time, insufficient knowledge and experience, it is inevitably needed to improve the quality of this book furtherly. We sincerely hope that readers and peers can give criticisms and corrections.

During the preparation of this book, Zhejiang University Sunnytech Co. Ltd. provides strong support for the development of chemical engineering principle experiment teaching software. The university and enterprise jointly develop the virtual laboratory of chemical engineering principle as the base of the first state-level chemical engineering virtual experiment center. This book provides teaching resources and cloud platform-Learning Class (www.walkclass.com). Readers can sweep the QR code in the back cover of the book to download Learning Class APP and join it, and then sweep the class QR code to obtain teaching resources on the mobile phone.

We express sincere gratitude to Zhejiang Supcon Instrument Co. Ltd. for offering some experimental teaching instruments. Also, we want to express sincere thanks to Dr. Satmon John who helps to check the English version of this book.

Moreover, we would like to extend our heartfelt thanks to all related schools, businesses and leaders providing strong support as well as people who are responsible for organizing, writing, modifying, reviewing, and printing the book.

Editor
May 1, 2017

Introduction

I Importance and purpose of chemical engineering principle experiment

Experimental class of chemical engineering principle is a course of chemical engineering and relevant majors with strong practicality. It is a practical course that studies the physical or chemical changes in material, design, and operations of equipment used in industrial processes under standard conditions. Any chemical or industrial process is composed of different types of unit operations. Therefore, in each experiment student will learn a unit operation, and be taught how to analyze and solve various engineering problems encountered in the unit operation. For science students, they can find out more about engineering principles, the method for measuring and testing experimental data with high accuracy and observe the relationship between complex technical process and mathematical model. Besides, they can realize that a seemingly complicated process can be illustrated and described with the basic principle.

Chemical engineering experiment principle is to teach the student and deepen their understanding of fundamental theories of chemical engineering. More importantly, it aims to train prospective students and technical workers to master basic experimental methods and skills as well as the ability to set up and perform experiments independently. Nevertheless, it lays a sound basis for undertaking scientific research and solving engineering problems in the future.

II Characteristics of chemical engineering principle experiment

Chemical engineering principle experiment involves numerous variables, materials and devices, facing complex practical engineering problems. Therefore, experimental methods would be different for varied objects. The course emphasizes practicality and engineering concepts and trains the ability of students throughout experiments. Based on the theories of chemical engineering, the course trains students to master experimental research methods and abilities to think, analyze and resolve problems independently.

III Teaching contents and methods of chemical engineering principle experiment

The teaching of chemical engineering principle experiment includes basic experimental knowledge and typical chemical unit operation experiments.

The teaching of basic experimental knowledge focuses on the purposes and requirements of chemical engineering principle experiment; equipment, process; experiment principle; method and procedure as well as experimental report.

The teaching of chemical unit operation experiments includes: comprehensive fluid mechanic experiment—fluid flow resistance experiment and centrifugal pump characteristic curve determination experiment; filtration experiment—constant pressure filtration and vacuum filtration; determination of convective heat transfer coefficient; packed column absorption; sieve-plate column rectification and efficiency determination; extraction column (rotating disc column/pulsed column) operation; determination of tunnel drying characteristic curve.

The teaching of chemical engineering principle experiment adopted the traditional method of combining theoretical lessons followed by an explanation of the experimental procedures. Traditional teaching method has obvious disadvantages.

① Many experimental devices are expensive characterized by complex operation, large space, high pollution, high risk (such as toxic and hazardous raw materials or reagents, high-temperature high-pressure or high-speed rotation, etc), high cost and energy consumption, which increase the difficulty in teaching practice.

② Teachers often arrange experiments for students after their theoretic teaching, but there is a great span between theory and actual experiment. In the conditions of complex operations, inadequate equipment and numerous students, students tend to be at a loss when conducting experiment due to less understanding and preparation, resulting in low efficiency and a waste of time and resources, and even the psychological burden to students.

③ Due to limited experimental equipment and location, students often perform equipment in batches with a large interval between experimental and theoretical teaching, which greatly increases the workload of teachers and reduces the effects of experiment.

Ⅳ Characteristics of virtual simulation chemical engineering experiment in this course

Virtual simulation experiment teaching in this course is undertaken by establishing highly simulated virtual experimental environments and objects based on the technologies of virtual reality, multimedia, human-computer interaction, database and network communication. The teaching of virtual simulation experiment of chemical engineering principle concentrates on science and engineering combination, interaction between teaching and research, connection between school and enterprise. It lays stress on various open experiment teaching and trainings centering on students. It is characterized by combined virtual simulation and practice and remote control, ensuring that students can receive the training of experimental skills and deepen the understanding of experiment process and knowledge by taking full advantages of virtual simulation experiment. Therefore, students can observe macro effects from micro experiments, understand research implications from experiments and combine experiments with practical production, these enhance the effects and achieve environmental protection and safety of chemical engineering experiment.

The introduction of virtual simulation experiment into the course of chemical engineering principle experiment has significant advantages.

① The shifting of experiment from actual operation to simulation training (virtual space) in a computer can help students make good preparations before class and significantly reduce the risks of experiment operations and save cost. Meanwhile, virtual simulation experiment software can increase the interest of students in operations and achieve the goal of joyful learning.

② Virtual simulation experiment has no limitations of time and space and can replace the actual experiment to avoid or greatly reduce the dangers posed by definite chemical experiments.

③ Virtual simulation experiment provides teachers with an optimized environment for their teaching. Rich and professional teaching resources in virtual simulation experiment provide a digital platform for teachers to prepare lessons. Teachers can design perfect teaching processes based on teaching objectives and students' needs, thus, strengthens the teaching effects in class.

④ Virtual simulation experiment creates a digital learning environment for students and it supports multiple learning modes such as autonomous learning and inquiry learning. In the digital learning environment, students can bring full play into their autonomy for inquiry, collaborative, independent and creative learning to develop their innovative spirit, collaboration spirit and practical abilities.

V New teaching method of chemical engineering principle experiment

1. Teaching method of integrating "theory, virtual simulation and actual practice"

In the new teaching mode of "theory, virtual simulation and actual practice", virtual simulation experiments are introduced after theoretical lecturing. Virtual simulation software can be easily used by students for preparation and practice prior to actual experiment. It builds a platform through which students can learn theories to guide practice, which is the most shining feature of the integrated teaching mode.

The teaching method of integrating "theory, virtual simulation and actual practice" in this course is described as follows.

① Prior to class, students are provided with theoretical study resources, virtual simulation software and learning tasks. Theoretical resources include experimental instructions, experimental precautions, how-to videos, etc. Through personalized learning, students can acquire relevant theoretical knowledge and understand the operations of experimental equipment.

② In the class, students are required to take part in theoretical tests and experimental evaluation using virtual simulation software to understand the mystery of theoretical knowledge and experimental operation. Only when students pass the theoretical tests especially virtual experimental evaluation can they conduct actual experiments.

③ Students are required to conduct actual experiments, observe experimental phenomena and record experimental data. As steps in operating virtual experiments and actual experiments are exactly the same, the safety problems in actual experiments can be well solved and experiments and theories can be deeply interpreted by students.

As mentioned above, the key to implement the integrated teaching mode of "theory, virtual simulation and actual practice" is to ensure that students can receive extracurricular theoretical knowledge and virtual simulation training as well as understand the fidelity between virtual simulation software and actual experiment. Therefore, virtual laboratory of chemical engineering principle and cloud platform are jointly developed by university and enterprise.

Cloud platform plays the roles of resource sharing, on-line learning, teaching management, efficient resource communication, task release, on-line testing and on-line virtual simulation software application, etc. Teaching based on cloud platform is a breakthrough in the traditional teaching way and it represents the spirit of sharing resources by everybody in the internet age.

The developed virtual simulation experiment software has the functions of teaching, practice and test. "Teaching" is the part through which students can have a preliminary understanding of experimental procedures by using video demonstration. "Practice" is the part through which students can repeatedly perform experimental practice for consolidation. "Test" is the part through which students can be tested on their mastery of experiments. The information of students' operations in this process is automatically recorded and the correctness of their operation steps is evaluated, which can help students check their knowledge by themselves; experiment data can also be automatically recorded and corresponding Excel forms can be generated, which make it eaisier for students to analyze data and write report.

Virtual simulation software also seamlessly integrates with cloud platform, i. e. it has the functions of starting from Web page, automatic data recording, on-line virtual experiment operation test and experiment report upload, making it much easier for virtual simulation experiments to be carried out and for the teachers to understand the learning situation of students anytime and anywhere.

The new experimental teaching mode of chemical engineering principle based on the integration of "theory, virtual simulation and actual practice" greatly eases the pressure on laboratory resources, saves teaching costs and improves teaching effects.

2. Teaching practice of integrating "theory, virtual simulation and actual practice"

When chemical engineering principle experiments are conducted in the teaching mode of "theory, virtual simulation and actual practice", teaching resources are uploaded to cloud platform first through which learning tasks and goals are released, then students learn and use relevant theoretical resources and perform virtual simulation chemical engineering experiments according to learning tasks and goals after receiving the notification. After students complete the practice and test of virtual simulation software for chemical engineering experiment, the software automatically saves operating procedures and data and rates related operational steps. Final data will be automatically submitted to the cloud platform. Teachers can log on the cloud platform to manage and make statistics of students' experimental preparations and operations. Students who pass the assessment can conduct actual experiments in a real laboratory. Experimental reports are also submitted to the cloud platform and examined by teachers online.

The teaching mode of integrating "theory, virtual simulation and actual practice" is popular among students. Virtual simulation chemical engineering software can assist students in making adequate preparations before a class by themselves. Experimental procedures through video teaching can help students make preparations and know laboratory equipment in advance. The software is a valuable tool for learning and can even inspire students and improve their understanding of experiment principles to resolve some problems. In addition, its realistic pictures and 3D roaming experience attract students. An accomplishment of preparations before class without guidance provides alternative teaching resources for students without experimental equipment.

Chapter 1 Research Methods of Chemical Engineering Principle Experiment

Chemical engineering principle is an engineering discipline. It aims to derive basic laws in the process and faces real and complex issues in production—specified material goes through a specific process in special equipment. The complexity of practical issue is not completely relevant to the process itself but involves complex geometric shapes and various physical properties of chemical engineering equipment. In addition to summarize production experience, experimental research is the basis for the establishment and development of chemical engineering discipline. On the basis of long-term experience and experimental research, evolved research methods of chemical engineering principle experiment mainly include direct experimental method, dimensional analysis method, mathematical model method and cold model experimental method.

I Direct experimental method

Direct experimental method means that experiments are directly conducted on objects by controlling or simulating some objective conditions to obtain relevant parameters and laws. It is the most basic way of solving practical engineering problems with direct effects and reliable results, but it is also greatly limited. The laws resulting from the direct experimental method only reflect the relationship between individual parameters but cannot reflect the whole nature of objects. Experiment results can only be applied to specific experimental conditions and equipment or extended to the phenomena with the same experimental conditions. In addition, experiments sometimes require large investment due to heavy workload, long time and great effort.

II Dimensional analysis method

Chemical engineering principle experiment faces an engineering issue affected by multiple variables, which requires experimental research methods to establish empirical relations to derive theoretical formulas. The laws of multiple variables affecting the process are often studied through experiments with grid method through which one of the variables is changed in sequence and other variables are fixed. If the number of variables is m, the number of conditions for each variable to change is n, then the number of experiments required is $n \times m$. If there are many involved variables, the number of experiments required is greatly increased, and then the workload of experiments must be heavy. In order to reduce the workload and make experimental results universal, dimensional analysis method is widely used to solve such problems of chemical engineering principle experiment.

III Mathematical model method

Mathematical model method means that objectives and content evaluation are defined with symbols, functional relations systematically and the change in relationships between each other is expressed with mathematical formulas. The expressed contents can be quantitative or qualitative but must be expressed in a quantitative manner. Therefore, mathematical model method tends to be operated in the quantitative manner. Basic characteristics: abstraction and simulation of evaluation; each parameter consists of factors associated with evaluation objects; the demonstration of the relationship between all relevant factors.

Mathematical model method is used according to the following steps based on adequate research.

① Rationally simplify complex problems without distortion to propose a physical model approximate to actual process and easily described with digital equation;

② Conduct mathematical description of the physical model to establish a mathematical model and then determine the initial and boundary conditions of the equation for solution;

③ Check the rationality of the mathematical model and measure its parameters through experiments.

IV Cold model experimental method

Cold model experiment is mainly used for physical simulation of flow state and transfer process. The results of experiment are analyzed and can be used to speculate real process. For example, the experimental study of gas-liquid mass transfer carried out with air, water and tracer, provides parameters for the design and renovation of gas-liquid mass transfer equipment; the experimental study of fluidization carried out with air and sand, provides the basis for the design of fluidized bed reactor. Therefore, the method through which the laws of various engineering factors affecting process are studied in the experiment equipment with similar structure and size to those of industrial equipment by using simulative materials of air, water and sand instead of real materials is called "cold model experiment".

Cold model experimental results can be applied to other fluids. The application of experimental results of small equipment to large industrial equipment can achieve the aim of "from one point to another" on material types and "from small to large" on equipment size. The relationships between physical quantities can be obtained through few direct and economic experiments combined with mathematical model method or dimensional analysis method, this method greatly reduces the workload in experiments. Analogue experiments that are improperly or impossibly performed under real conditions can be carried out to reduce the risks arising from experiments. However, cold model experimental results can be used for the design and development of industrial processes only after they are corrected with chemical reactions features.

Chapter 2 Processing of Experimental Data

Plenty of raw data measured are the main results of experiments, after calculation, the final results need to be derived into empirical formulas or expressed in chart for analysis. However, there are always some errors in the data introduced by measuring instruments, operation methods and human observations during the process of experiment, so the reliability of experimental data should be assessed in an objective manner when the original data are organized.

I Error analysis of experimental data

The purpose of error analysis is to evaluate the accuracy of experimental data. Through error analysis, error sources and their impacts can be identified, invalid components contained in the data can be removed, main factors affecting the accuracy of experiement can be observed, and experimental plan can be further revised to improve the accuracy of experiment. Therefore, analysis and estimation of experimental errors are of great significance for the evaluation of experimental results and plan design.

1. True value and average value

True value, theoretical value or defined value, refers to the value of determining the objective existence of a physical quantity to be measured. Generally, the true value of a physical quantity is unknown and required to be measured. Strictly speaking, true value cannot be measured due to imperfect measuring instruments, measurement methods, environment, human observation and measurement procedures. The true value in scientific experiments can be defined as follows: in case of unlimited experimental measurements, the occurrence probability of positive or negative errors is equal according to the distribution law of error. Then systematic errors are eliminated in a detailed manner and measured values are averaged thus the value extremely close to the truth can be obtained. However, the number of measurements in the experiment is always limited in practice. The average value obtained through limited measurements can only be the value approximate to true value. The commonly used average values are as follows.

(1) **Arithmetic average**

Arithmetic average is the most often used average value.

Assuming that $x_1, x_2, \ldots, x_n$ are various measured values and n represents the number of measurements, then the arithmetic average value is

$$\bar{x} = \frac{x_1 + x_2 + \cdots + x_n}{n} = \frac{\sum_{i=1}^{n} x_i}{n} \tag{2-1}$$

(2) Geometric average

Geometric average is the average obtained by successively multiplying a set of n measured values and finding the n-th root. That is

$$\bar{x}_n = \sqrt[n]{x_1 x_2 \cdots x_n} \tag{2-2}$$

(3) Root mean square average

$$\bar{x}_s = \sqrt{\frac{x_1^2 + x_2^2 + \cdots + x_n^2}{n}} = \sqrt{\frac{\sum_{i=1}^{n} x_i^2}{n}} \tag{2-3}$$

(4) Logarithmic average

The distribution curves of chemical reaction, heat and mass transfer mostly have the properties of logarithm. In this case, logarithmic average is commonly used for characterization.

The logarithmic average of assumptive value x_1 and x_2 is

$$\bar{x}_m = \frac{x_1 - x_2}{\ln x_1 - \ln x_2} = \frac{x_1 - x_2}{\ln \frac{x_1}{x_2}} \tag{2-4}$$

when $x_1/x_2 = 2$, $\bar{x}_m = 1.443$, $\bar{x} = 1.50$, $(\bar{x}_m - \bar{x})/\bar{x}_m = 4.2\%$, means $x_1/x_2 \leqslant 2$ and the resulting error does not exceed 4.2%.

It should be noted that the logarithmic average of a variable is always less than its arithmetic average. When x_1/x_2 is less than or equal to 2, the arithmetic average can be replaced by logarithmic average.

(5) Weighted average

Assuming that the same physical quantity is measured with different methods or by different people, that is, the reliability of each measured value through unequal precision measurement is not the same. In the calculation of an average, reliable values are often calculated by weighted average method.

$$w = \frac{w_1 x_1 + w_2 x_2 + \cdots + w_n x_n}{w_1 + w_2 + \cdots + w_n} = \frac{\sum w_i x_i}{\sum w_i} \tag{2-5}$$

where, $w_1, w_2, \cdots, w_n$ represent the corresponding weight of each observed value. The weight of each observed value is generally determined by experience.

(6) Median

It refers to the intermediate value when a set of observed values are arranged in a certain order. If the number of observations is even, the median is the average of the two values in middle. The greatest advantage of the median is that it is easily calculated independent of change in both ends. Median is a type of order statistics in terms of design. Only when the distribution of observed values is normal can it represent the best value for a set of observations.

The introduction of the above average values aims to find the value approximate to the true value out of a set of measured values. In chemical engineering experiments and scientific research, most data are normally distributed, so arithmetic average is typically used.

2. Error nature and classification

Regardless of precise instruments, perfect methods and careful experimenters during any measurement, the results measured at different time may not be consistent but have some errors and deviations. Strictly speaking, error refers to the difference between the measured value (including the values directly and indirectly measured) and the true value (objective accurate value). Deviation refers to the difference between the measured value and average. However, people often confuse them and make no distinction. According to the nature and causes of error, error can be classified into systematic error, accidental error and gross error.

① Systematic error (constant error). Under the same experimental condition, when the same quantity is measured for several times, the error whose value and positive-negative direction always maintain constant or change with the experimental conditions according to some rule is called systematic error. For example, measuring instruments with inaccurate scale and non-calibrated zero; changes in experimental environment, such as outside temperature, pressure and humidity; experimenter habits and inclinations.

Systematic error is the main part of measurement errors, so the elimination and estimation of systematic error are significant to improve the measurement accuracy. Generally, the occurrence of system errors is uniform. The causes of their occurrence are often observable or can be eliminated through careful calibration or inspection.

② Random error (accidental error). Under the same condition, when the same physical quantity is measured, the absolute values of error are changeable. The error is either positive or negative without certain laws and unpredictable but entirely subject to the statistical law. When the same physical quantity is measured for several times, with the increase in the number of measurements, the arithmetic average of the random error is usually zero. Therefore, the arithmetic average measured by multiple times would be approximate to the true value.

③ Gross error. It refers to the error caused by wrong operations or human faults. Such errors are often greatly different from normal values and should be eliminated at the time of data analysis.

3. Error representation

There are always errors in the measurement when using any measuring tools or instruments. The measured value might not be exactly equal to the true value but approximate to it. Measurement quality is conditional on measurement accuracy. Measurement accuracy is estimated according to the value of the measurement error. The smaller the error of measurement result is, the more accurate the measurement is.

Measurement error includes the error of measured points and the error of the measured series (set). They are differently represented.

(1) The representation of error of measured points

① Absolute error D. The absolute value of the difference between the measured value x and the true value A_0 in the measurement set is called absolute error. It is represented as follows

$$D = |x - A_0| \qquad (2\text{-}6)$$

That is
$$x - A_0 = \pm D \quad x - D \leqslant A_0 \leqslant x + D \qquad (2\text{-}7)$$

Since the true value A_0 cannot be obtained, the above equations are only significant in theory. The indication of a higher level standard instrument is often used as the actual value A to replace the true value A_0. Due to the small errors existed in higher level standard instrument, A is not equal to A_0 but more approximate to A_0 than x. The difference between x and A is called indicative absolute error. It is represented as follows

$$d = |x - A| \qquad (2\text{-}8)$$

The number opposite to d is called corrected value, represented as

$$C = -d \qquad (2\text{-}9)$$

The corrected value of tested instrument C can be given by higher level standard equipment through testing. The actual value A of the instrument can be calculated with the corrected value.

② Relative error. Relative error is generally used to judge the accuracy of a measured value. The ratio between the absolute error of indication d and actual measured value A is called actual relative error. It is represented as follows

$$E_r(x) = \frac{d}{A} \times 100\% \qquad (2\text{-}10)$$

Relative error of measuring instrument indication x in place of the actual value A is called relative error of indication. It is represented as follows

$$E_r(x) = \frac{d}{x} \times 100\% \qquad (2\text{-}11)$$

Generally, the use of relative error of indication is more applicable except theoretical analysis.

③ Fiducial error. In order to calculate and divide the accuracy grade of instruments, the concept of fiducial error is put forward. It is the ratio between absolute error of instrument indication and measuring range.

$$\delta_A = \frac{\text{absolute error of indication}}{\text{range}} \times 100\% = \frac{d}{X_n} \times 100\% \qquad (2\text{-}12)$$

where d — absolute error of indication;

X_n — upper limit of scale — lower limit of scale.

(2) The representation of error of measuring series (set)

① Range error. Range error refers to the difference between the maximum value and the minimum value within a set of measurements to indicate the scope of error change. The concept of error coefficient is often used.

$$K_1 = \frac{L}{\bar{x}} \tag{2-13}$$

where K_1—maximum error coefficient;

L—range error (the difference between the maximum value and the minimum value within a set of measurements);

$\bar{x}$—arithmetic average.

The biggest disadvantage of range error is that K_1 only depends on both extreme values but is irrelevant to the number of measurements.

② Arithmetic mean error. Arithmetic mean error is the average of errors at each measuring point.

$$\delta_{avg} = \frac{\sum_{i=1}^{n} |d_i|}{n} \tag{2-14}$$

where n—number of measurements;

d—deviation between the measured value and the mean value, $d_i = x_i - \bar{x}$.

Arithmetic mean error is a good way to represent errors but unable to reflect the correspondence of errors between various measurements.

③ Standard error. Standard error is also known as the root mean square error.

$$\sigma = \sqrt{\frac{\sum_{i=1}^{n} d_i^2}{n}} \tag{2-15}$$

Standard error is sensitive to bigger or smaller errors in a set of measurements, so it is a better way to reflect accuracy.

Equation (2-15) is applied to unlimited measurements. In practice, the number of measurements is limited, so equation (2-16) is used

$$\sigma = \sqrt{\frac{\sum_{i=1}^{n} d_i^2}{n-1}} \tag{2-16}$$

Standard error is not a specific error. The value of σ only indicates the dispersion of each observed value belonging to the collection of equally accurate measurements to their arithmetic mean values under certain conditions. If the value of σ is smaller, the dispersion of each measured value to their arithmetic mean values is lower and the measurement accuracy is higher, otherwise the accuracy is lower.

In chemical engineering principle experiment, the minimum scale values of most commonly used U-tube differential pressure meter, rotameter, stopwatch, measuring cylinder and voltage meter are regarded as their maximum errors in principle and the half of the minimum scale values are regarded as their absolute error values.

④ Probable error. Probable error is also known as the probable deviation represented by the symbol γ. It means that if plus and minus sign are ignored in a set of measurements, the observed values with an error greater than γ or less than γ account for 50% of observations respectively.

$$\gamma = 0.6745\sigma \tag{2-17}$$

Probable error has been gradually replaced by the standard error in recent years.

Regardless of comparing the accuracy of measured values or evaluating the quality of measurement results, relative error and standard error are preferred in the representation of errors. Standard error is often used in literature.

4. Precision, correctness and accuracy of experimental data

Students are frequently satisfied with the reproducibility of experimental data but ignore the accuracy of measured values. Absolute true values are unknowable, so people can only develop some international standards as the references for measuring the accuracy of instruments. The terms in domestic and international literatures are not identical. The use of precision, correctness and accuracy has always been a confusing problem. Recently most people tend to accept the following expressions.

① Precision. The degree of reproducibility of measured values in the measurement. It reflects the influence of accidental errors. High precision means small accidental error.

② Correctness. Comprehensive degree of all systematic errors in the measurement under specified conditions.

③ Accuracy (degree of accuracy). Deviation between the measured value and the true value. It reflects the composite error involving systematic error and random error.

To illustrate the difference between precision and accuracy, the following example of target shooting can be used.

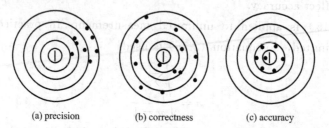

(a) precision　　　(b) correctness　　　(c) accuracy

Figure 2-1　The relationship of precision, correctness and accuracy

As shown in Figure 2-1, (b) has small systematic errors but big random errors, which means the correctness is high but the precision is low; (a) has big systematic errors but small random errors, which means the correctness is low but the precision is high; (c) has small systematic and random errors indicating that the degree of accuracy is high. In practical measurement, there is no definite true value like target, people should manage to determine the unknown true value.

High accuracy of experiment or measurement does not mean high correctness. In addition, high correctness does not mean high precision. However, if accuracy is high, precision and correctness must be high.

Absolute true values are unknowable, with the development of human understanding, people can obtain the value more approximate to the absolute true value.

5. Accuracy and measurement error of instrument

(1) Accuracy and measurement error of electrical instruments

The accuracy of the instruments is often represented with the maximum fiducial error and the accuracy class of instrument. Maximum fiducial error of instrument is defined as follows

$$\text{Maximum fiducial error} = \frac{\text{absolute error of instrument indication}}{\text{absolute value in relevant range of the instrument}} \times 100\% \quad (2\text{-}18)$$

The absolute error of instrument indication in equation (2-18) refers to the maximum absolute value of the difference between the measured value and standard value of the measured parameter. As for multi-range instruments, the absolute error and range of different indications are not the same.

Equation (2-18) shows that if the absolute errors of instrument indication are equal, the larger the measuring range is, the smaller the maximum fiducial error is.

The accuracy of electrical instruments is classified into seven levels in China: 0.1, 0.2, 0.5, 1.0, 1.5, 2.5 and 5.0. If the accuracy level of an instrument is 2.5, it indicates the maximum fiducial error of this instrument is 2.5%.

How to estimate the absolute and relative errors of measurement when using an instrument?

Assuming that the accuracy level of the instrument is P, the maximum fiducial error is $P\%$, the measurement range of instrument is x_n and the measured value of the instrument is x_i, then the error of the indication obtained from equation (2-18) is

$$\text{Absolute error } D \leqslant x_n \times P\% \quad (2\text{-}19)$$

$$\text{Relative error } E_r(x) = \frac{D}{x_i} \leqslant \frac{x_n}{x_i} \times P\% \quad (2\text{-}20)$$

Equation (2-20) shows that

① If the grade of accuracy P and the measurement range x_n of the instrument are fixed, the larger the measured indication x_i is, the smaller the measured relative error is.

② When choosing an instrument, the accuracy level of the instrument is not the only standard, because the relative error of measurement is related to $\frac{x_n}{x_i}$. The accuracy class and $\frac{x_n}{x_i}$ of the instrument should be taken into account.

(2) Relationship between accuracy of instruments like balance and their measurement

The accuracy of these instruments is represented by equation (2-21)

$$\text{Accuracy} = \frac{\text{nominal division value}}{\text{range}} \quad (2\text{-}21)$$

The nominal division value refers to the minimum division unit of readings which are surely accurate in the measurement, i.e. the numerical value of each minimum division. For example, the nominal division value (sensitive quantity) of TG-328A balance is 0.1 mg and its measuring range is 0~200 g, then

$$\text{Accuracy} = \frac{0.1}{(200-0) \times 10^3} = 5 \times 10^{-7} \quad (2\text{-}22)$$

If the accuracy of the instrument is known, the nominal division value can also be obtained from equation (2-21).

When using these instruments, the errors of measured values can be determined with the following formula

$$\begin{cases} \text{absolute error} \leqslant \text{nominal division value} \\ \text{relative error} \leqslant \dfrac{\text{nominal division value}}{\text{measured value}} \end{cases}$$

(3) Actual error of measured value

According to the accuracy of instrument, the measurement errors determined with the above method are generally much smaller than the actual errors of measured value. There are numerous factors to cause errors. For example, instrument has not been properly adjusted. Improper vertical, horizontal and zero positions after adjustment would cause errors; non-compliance of actual work conditions of instrument with specified normal conditions would cause additional errors; parts wear and the changes in assembly of instrument after it has been used for a long time would also cause errors; personal habits in measurement may cause errors; signals received by instrument may not be similar to the measured value resulting in errors; instrument circuits may be disturbed to cause errors…

In other word, there are many factors to cause actual errors of measured values. In order to obtain accurate measurements, good instruments, scientific style and methods as well as solid theoretical knowledge and practical experience are required.

6. Significant figure and algorithm

In science and engineering, a number of certain digits are always used to represent a measured or calculated value. The accuracy of a numerical value cannot be determined by the number of digits after the decimal point. The number of digits of a measured value in experiment is limited. The last digit of the value is often estimated by the accuracy of instrument. That is, the accuracy of measuring instrument should be generally indicated by one tenth of its smallest scale. The accuracy of numerical value is determined by the digits of significant figure.

The representation of directly measured or calculated results in experiment by a number of certain digits is significant. Students often come up with these two ideas: the more the number of digits after the decimal point of a numerical value is, the more accurate the numerical value is, or the more the number of digits is reserved in a calculated result, the more accurate the calculated result is. In fact, both ideas are incorrect. There are two reasons. First, accuracy is determined by the unit of physical quantity rather than the position of the decimal point. For example, the accuracy is the same if the electromotive force of a thermocouple measured with potentiometer is 764.9 μV or 0.7649mV; second, measuring instruments have limited precision (or sensitivity). Taking the above example for description, the precision of the potentiometer can only be 0.1 μV or 0.0001mV. The accuracy of the result through calculation would never exceed the error limit allowed by the instrument.

Thus it can be seen that the representation of a measured or calculated value by a number of certain digits depends on the precision of measuring instrument. The accuracy of numerical value is determined by the digits of significant figure. In the above example, the precision of numerical value is 0.1μV and the accuracy is represented by a significant figure of four digits.

(1) Concept of significant figure

In addition to the positioning number "0", the other numbers of data are effective. For example, 0.0037 has only two significant figures but 370.0 has four significant figures. It should be emphasized that significant figures may not be reliable.

① Significant figure of directly measured data. The data measured in experiment are only approximate values. In the measurement, a value can generally be read out by one digit after the minimum mark of instrument. The last number is the estimate included in effective numbers. For example, the minimum mark of Grade Ⅱ thermometer is 0.1℃. The measured value can be read out to 0.01℃, such as 15.16℃. The value has four significant figures but only three figures are reliable. The last one is not reliable and called dubious number. Only one dubious number is kept when the measured values are recorded. The value of 15.2℃ should be recorded as 15.20℃, indicating that it has four significant figures.

In order to clearly represent the accuracy of numerical values and display the digits of significant figure, index is often used, that is, numerical values are represented by the product of a decimal and its integer power of 10. The numeration with integer powers of 10 is called scientific notation.

For example, 75200 is recorded as 7.520×10^4 when it has 4 significant digits.

75200 is recorded as 7.52×10^4 when it has 3 significant digits.

75200 is recorded as 7.5×10^4 when it has 2 significant digits.

0.00478 is recorded as 4.780×10^{-3} when it has 4 significant digits.

0.00478 is recorded as 4.78×10^{-3} when it has 3 significant digits.

0.00478 is recorded as 4.8×10^{-3} when it has 2 significant digits.

The significant digits of measurement depend on the requirement for the accuracy of results and the accuracy of measuring instrument itself.

② Significant figure of indirectly measured data. In experiment, in addition to the above numbers with units, other constants without units like π, e and certain factors like $\sqrt{2}$ are often used. The number of their significant digits can be infinite. Their significant digits depend on the digits of significant figure of original data employed in calculation. Assuming that the sizable number of the maximum digits has n digits, the digits of the above constants for reference should be $n+2$ to avoid larger errors when using the data.

In the data calculation, all intermediate results can be read out by two digits more than the original data with maximum significant figures to obtain higher accuracy of calculation. However, it is better for intermediate results to get more significant digits in the regression analysis and calculation. Significant figure should have at least 6 digits to weaken the quick accumulation of round-off errors.

The data to represent the value of error should generally have two significant figures.

(2) Calculation of significant figure

The accuracy of calculated results should not be higher than that of original data, so more digits preserved in a data do not necessarily improve the accuracy of the calculation, but it is a waste of time. However, few digits preserved in data would reduce the precision. The choice of the number of digits in calculation is determined according to the operation rules of significant figure.

① In the addition and subtraction, the number of digits after the decimal point retained in calculation results should be the same as that of the number with the fewest digits after the decimal point. For example, when 13.65, 0.0082 and 1.632 are added, they should be recorded as

$$13.65 + 0.01 + 1.63 = 15.29$$

In the addition, the number of digits after the decimal point is included in and the unknown number is substituted into x, then the equation obtained is

$$\begin{array}{r} 1\,3\,.\,6\,5\,x\,x \\ 0\,.\,0\,0\,8\,2 \\ +\ \ 1\,.\,6\,3\,2\,x \\ \hline 1\,5\,.\,2\,9\,x\,x \end{array}$$

Thus 15.29 is the most logical result.

② In the multiplication and division, the number of significant digits of calculation results should be the same as that of the original number with the least number of significant digits.

For example, the result of $1.3048 \times 236 = 307.9328$ is rounded to 308 based on the number of 236.

③ In the power and extraction of a root, the number of significant digits of calculation results should be the same as that of its base.

④ In the logarithm calculation, the number of significant digits of logarithm should be the same as that of antilogarithm (does not include the positioning part).

⑤ In calculation, when the first significant figure is equal to or greater than 8, the number of significant digits can be increased by one digit. For example, 8.13 only has three digits but it can be regarded as a number with four significant digits in calculation.

(3) Rounding principles

When the experiment result has too many digits due to calculation or other reasons, the result needs to be rounded to required number of digits. It is better to use the rounding principles as follows.

① When the leftmost digit of the number to be rounded is less than 5, it should be rounded, i. e. the retained digits are unchanged.

② When the leftmost digit of the number to be rounded is greater than 5 or equal to 5 followed by the number other than 0, the digit should be rounded up by 1, i. e. the retained last digit is increased by 1.

③ When the leftmost digit of the number to be rounded is 5 and there is no digit or the digit is 0 on the right of 5, if the retained last digit is odd, the digit should be rounded up by 1, if the retained last digit is even or 0, the digit should be rounded down, i. e. "up for odd and down for even".

The rule is also called even rule of number rounding, that is, "Banker's Rounding".

For example: 2. 8635 is recorded as 2. 864 when four significant figures are retained.
                   2. 8635 is recorded as 2. 86 when three significant figures are retained.
                   2. 8665 is recorded as 2. 866 when four significant figures are retained.
                   2. 8665 is recorded as 2. 87 when three significant figures are retained.
                   2. 866501 is recorded as 2. 867 when four significant figures are retained.
                   2. 86499 is recorded as 2. 86 when three significant figures are retained.

7. Blunder error rounding

Sometimes a few large or small values present in the measurement. These abnormal values have a bad influence on measuring results and should be discarded. However, it is not necessary to achieve the consistency of results by discarding the "bad values" at will. Because the arbitrary discarding of some normal measured values with large errors would produce false results.

How to judge whether the measured values are abnormal? The simplest method is the triple standard error principle.

Based on the theory of probability, the probability of error greater than 3σ (mean square error) is only 0.3%, so the value is often called limit error, that is

$$\delta_{\text{limit}} = 3\sigma \tag{2-23}$$

If the error of individual measurement exceeds 3σ, then it can be considered as blunder error and discarded. It is important to find the way to discard the questionable value in a few measured values. The theory of probability is not applicable due to a few measurements and individual abnormal measurements have great impact on the arithmetic average.

Someone proposed a simple judgment method through which questionable measured values are discarded, the average and average error δ of remaining measured values are calculated, then the deviation d between questionable measured values and average can be calculated. If $d \geqslant 4\delta$, the probability of questionable value is only about one in a thousand.

8. Error propagation in indirect measurement

In many experiments and researches, the desired results are not directly measured with instruments but are calculated by substituting some directly measured values into certain theoretical relations, i. e. indirectly measured values. Since directly measured values always have some errors, they would inevitably cause some errors in indirect measurements, i. e. the errors of directly measured values are inevitably propagated to the indirect measured values resulting in errors.

Error propagation formula: when there is a functional relationship between indirectly measured value (y) and directly measured value ($x_1, x_2, \cdots, x_n$), that is

$$y = f(x_1, x_2, \cdots, x_n) \tag{2-24}$$

The differential formula is

$$dy = \frac{\partial y}{\partial x_1} dx_1 + \frac{\partial y}{\partial x_2} dx_2 + \cdots + \frac{\partial y}{\partial x_n} dx_n \tag{2-25}$$

$$\frac{dy}{y} = \frac{1}{f(x_1, x_2, \cdots, x_n)} \left(\frac{\partial y}{\partial x_1} dx_1 + \frac{\partial y}{\partial x_2} dx_2 + \cdots + \frac{\partial y}{\partial x_n} dx_n \right) \tag{2-26}$$

According to the two formulas, when the error of directly measured value ($\Delta x_1, \Delta x_2, \cdots, \Delta x_n$) is small and the worst situation is taken into account, the absolute value can be obtained by the cumulative summation of individual errors, the error of indirectly measured value (Δy and $\frac{\Delta y}{y}$) can be obtained

$$\Delta y = \left| \frac{\partial y}{\partial x_1} \right| \cdot |dx_1| + \left| \frac{\partial y}{\partial x_2} \right| \cdot |dx_2| + \cdots + \left| \frac{\partial y}{\partial x_n} \right| \cdot |dx_n| \tag{2-27}$$

$$E_r(y) = \frac{\Delta y}{y} = \frac{1}{f(x_1, x_2, \cdots, x_n)} \left(\left| \frac{\partial y}{\partial x_1} \right| \cdot |dx_1| + \left| \frac{\partial y}{\partial x_2} \right| \cdot |dx_2| + \cdots + \left| \frac{\partial y}{\partial x_n} \right| \cdot |dx_n| \right) \tag{2-28}$$

The two equations are the error propagation formulas to calculate the error of indirectly measured value with the error of directly measured value.

The propagation formula of standard error is

$$\sigma_y = \sqrt{\left(\frac{\partial y}{\partial x_1} \right)^2 \sigma_{x_1}^2 + \left(\frac{\partial y}{\partial x_2} \right)^2 \sigma_{x_2}^2 + \cdots + \left(\frac{\partial y}{\partial x_n} \right)^2 \sigma_{x_n}^2} \tag{2-29}$$

where, σ_{x_1} and σ_{x_2} are the standard errors of directly measured value and σ_y is the standard error of indirectly measured value.

The functional expressions to calculate errors are listed in Table 2-1.

Table 2-1 Functional expressions of error propagation formula

Functional expression	Error propagation formula	
	Maximum absolute error Δy	Maximum relative error $E_r(y)$
$y = x_1 + x_2 + \cdots + x_n$	$\Delta y = \pm(|\Delta x_1| + |\Delta x_2| + \cdots + |\Delta x_n|)$	$E_r(y) = \Delta y / y$
$y = x_1 - x_2$	$\Delta y = \pm(|\Delta x_1| + |\Delta x_2|)$	$E_r(y) = \Delta y / y$
$y = x_1 x_2$	$\Delta y = \pm(|x_1 \Delta x_2| + |x_2 \Delta x_1|)$	$E_r(y) = \pm \left(\left\| \frac{\Delta x_1}{x_1} \right\| + \left\| \frac{\Delta x_2}{x_2} \right\| \right)$
$y = x_1 x_2 x_3$	$\Delta y = \pm(|x_1 x_2 \Delta x_3| + |x_1 x_3 \Delta x_2| + |x_2 x_3 \Delta x_1|)$	$E_r(y) = \pm \left(\left\| \frac{\Delta x_1}{x_1} \right\| + \left\| \frac{\Delta x_2}{x_2} \right\| + \left\| \frac{\Delta x_3}{x_3} \right\| \right)$
$y = x^n$	$\Delta y = \pm(|nx^{n-1} \Delta x|)$	$E_r(y) = \pm \left(n \left\| \frac{\Delta x}{x} \right\| \right)$

Continued

Functional expression	Error propagation formula	
	Maximum absolute error Δy	Maximum relative error $E_r(y)$
$y = \sqrt[n]{x}$	$\Delta y = \pm \left(\left\| \dfrac{1}{n} x^{\frac{1}{n}-1} \Delta x \right\| \right)$	$E_r(y) = \pm \left(\dfrac{1}{n} \left\| \dfrac{\Delta x}{x} \right\| \right)$
$y = x_1 / x_2$	$\Delta y = \pm \left(\dfrac{\|x_2 \Delta x_1\| + \|x_1 \Delta x_2\|}{x_2^2} \right)$	$E_r(y) = \pm \left(\left\| \dfrac{\Delta x_1}{x_1} \right\| + \left\| \dfrac{\Delta x_2}{x_2} \right\| \right)$
$y = cx$	$\Delta y = \pm \|c \Delta x\|$	$E_r(y) = \pm \left(\left\| \dfrac{\Delta x}{x} \right\| \right)$
$y = \lg x$	$\Delta y = \pm \left\| 0.4343 \dfrac{\Delta x}{x} \right\|$	$E_r(y) = \Delta y / y$
$y = \ln x$	$\Delta y = \pm \left\| \dfrac{\Delta x}{x} \right\|$	$E_r(y) = \Delta y / y$

II Processing of experimental data

The processing of experimental data refers to the deduction of true value of a measured value to draw the conclusions of certain laws by means of measurement and mathematical computation based on study concept and status. Therefore, the processing of experimental data can enable people to observe the quantitative relationships between variables and further analyze experimental phenomena to discover laws for the guidance of production and design.

1. Tabulation method

For easy and orderly processing of data, experimental data are required to be organized. Record and calculation tables are required to be designed in advance according to the contents of the experiment.

Experimental data record table generally consists of the following columns: test data column (raw data), intermediate calculation column and experimental result column. The contents depend on the experiment.

Attention should be paid to the preparation of record tables as follows.

① Each column must be titled with names and units of physical quantities. Names should be indicated with symbols. Units and orders of magnitude should be written in the title bar of each symbol.

② Digits to be recorded should be limited to significant figures. The digits of sizable numbers in the same column should be equal.

③ The numbers with large or small orders of magnitude can be recorded as integers in the data bar when they are multiplied by appropriate times in the title bar.

For example: $Re = 25000 = 2.5 \times 10^4$

The number recorded as $Re \times 10^{-4}$ in the title bar can be recorded as 2.5 in the data bar.

④ Data should be arranged in the order of measured data, intermediate calculated data and experimental results from left to right for easy record, calculation and organization. The columns of measured data, intermediate calculated data and experimental results can also be kept separated from each other.

To simplify calculation results, save time and reduce errors, constant induction method should be used as much as possible to calculate constant terms in advance.

2. Graphic method

The regularities of data are generally not easily discovered with the above method. If the data of independent and dependent variables in the table of experimental results are marked on a graph paper, the relationships between independent and dependent variables are indicated with graphs, a simple and intuitive view is presented and the regularities of results are compared and showed easily.

The basic principles for correct graphics in engineering experiment must be followed.

① In the coordinate system of two variables, the horizontal axis always represents independent variables and the vertical axis always represents dependent variables. Names, symbols and units of variables should be marked at the side of both axes. Beginners should pay particular attention to units as they often ignore them under the influence of pure numbers.

② The use of coordinate division should reflect the digits of significant figures of experimental data. That is, it should be consistent with the precision of marked values and easily read. Coordinate index values are not necessarily from zero and graphics should fill the whole coordinate system.

Coordinate division means the size of numerical values on each coordinate along the x and y axis. Actually, it refers to the selection of coordinate scale. Completely different conclusions would be drawn from the plotted graphics for the same set of data due to different selected scales. For example, when the functional relationship between x and y is determined, a set of experimental data are obtained as Table 2-2.

Table 2-2　Experimental data

x	1.00	2.00	3.00	4.00
y	8.0	8.2	8.3	8.0

The data is marked in Figure 2-2. The scale of y-coordinate in Figure 2-2(a) is different from that in Figure 2-2(b) resulting in completely different forms. Based on the curves in Figure 2-2(a), y is irrelevant to x or y is a constant. Figure 2-2(b) shows that y changes with x. From these two graphs, it seems that the selected coordinate scale can determine the nature of functions. In fact, it is not true. Mathematically, functional relationship only depends on the values of independent and dependent variables. Properties of function is decided by its nature. Improper selection of scale does not reveal the internal rule of function and cannot change the nature of function. In Figure 2-2(a) and (b), the scale of Figure 2-2(a) does not fit the experiment accuracy and therefore the change rule of y, x cannot be described, so the accuracy of experiment must be considered when the scale is selected.

If the experimental accuracy of y and x is respectively Δy and Δx and points fluctuate in the rectangle $2\Delta y$ and $2\Delta x$, the correct plotting method should make

$$m_x \times 2\Delta x = m_y \times 2\Delta y \qquad (2\text{-}30)$$

where, m_x, m_y is respectively the scale coefficient of x and y, that is, millimeters of each scale (mm/unit).

In Figure 2-2, if the accuracy of y and x is respectively $\Delta y = \pm 0.2$ and $\Delta x = \pm 0.05$

$$\frac{\Delta y}{\Delta x} = \frac{0.2}{0.05} = 4.0 \qquad (2\text{-}31)$$

if $\qquad m_x = 4.0 (\text{mm/unit})$
Then $\qquad m_y = 1.0 (\text{mm/unit})$

③ If they are on the same coordinate system, while fitting a few sets of measured values, different symbol should be used for each set (for example: •、×、△) to make difference. If n sets of

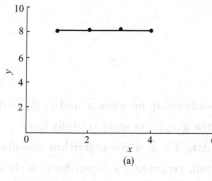
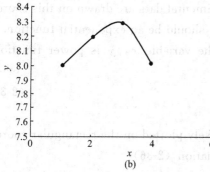

Figure 2-2 Functional relation of $y = f(x)$

different functions are plotted on the same coordinate system, functional relationship or name should be marked on the curves or direction arrows of readings should be indicated.

④ Straight lines are most easily drawn and used when fitting experimental curves. Therefore, curved lines should be straightened as much as possible in the processing of data. To this end, variables should be changed or different coordinate system should be selected according to different conditions. For example, single-logarithm and double-logarithm coordinate system are often used in the processing of chemical engineering experimental data. In addition, it is desired that the slope of the obtained curve should be performed with a scale division approximate to 1.

Straight line graphs of the following functions can be obtained on a different coordinate system.

① Linear function. The functional relationship between the variable x, y is

$$y = a + bx \qquad (2\text{-}32)$$

② Exponential function. If the relationship between the variable x, y is exponential function, then

$$y = k e^{mx} \qquad (2\text{-}33)$$

where k, m represent undetermined coefficients.

In this case, if the data of x, y are graphed in rectangular coordinate system, a curve shall display on the graph. If taking logarithm on both sides
Then $\qquad \lg y = \lg k + mx \lg e \qquad (2\text{-}34)$

Let
$$\lg y = Y$$
$$m \lg e = b_1$$
$$\lg k = a_1$$

Then the above equation is converted into
$$Y = a_1 + b_1 x \tag{2-35}$$

Through the above processing, there is a linear relationship between x and Y. By fitting $\lg y = Y$ and x on the rectangular coordinate system, the graph presents straight line.

To avoid taking logarithm of each experimental data Y, a single-logarithm coordinate system can be utilized. One axis of the coordinate system represents a logarithmic scale and the other still represents rectangular scale. If the experimental data are drawn on this coordinate system and straight line presents, the correlation should be an exponential function.

③ Power function. If the relationship between the variable x, y is power function, then
$$y = k x^m \tag{2-36}$$

where, k, m represent undetermined coefficients.

There must be a curve if equation (2-36) is precisely plotted on the rectangular coordinate system. By taking logarithms on both sides of equation (2-36)
$$\lg y = \lg k + m \lg x \tag{2-37}$$

Let
$$\lg y = Y$$
$$\lg x = X$$
$$m = b_2$$
$$\lg k = a_2$$

Then equation (2-37) is converted into
$$Y = a_2 + b_2 X \tag{2-38}$$

According to equation (2-38), take logarithms $\lg y = Y$ and $\lg x = X$ of experimental data x, y, there is a straight line when they are plotted on the rectangular coordinate system. Similarly, to resolve the trouble of taking logarithm every time, x, y can also be directly plotted on the double-logarithm coordinate system and the results are equal. Things need to be paid attention to when double-logarithm coordinate system is used.

a. The number marked on the logarithmic coordinate system is true value rather than logarithm.

b. As $\lg 1 = 0$, a straight line of $x = 0$ on the ordinary coordinate system is the straight line of $x = 1$ on the logarithmic coordinate system.

c. It should be noted that when determining the index m and coefficient k, m cannot be calculated with the method used in ordinary coordinate, but
$$m = \frac{\text{distance between two points on the } y \text{ axis}}{\text{distance between two points on the } x \text{ axis}} \neq \frac{\text{difference between readings of } y \text{ axis}}{\text{difference between readings of } x \text{ axis}}$$

k is the value of y at the intersection point of the line and $x = 1$, or it can be obtained by substituting any set of (x_1, y_1) values on the straight line into equation (2-36).

The coefficient k, m can also be calculated with the average method or the least square method.

Reasonable methods should be used to fit experimental points into smooth curves. Discrete experimental points should be roughly averaged for fitting. It is obvious that the results have large errors and low accuracy.

It is best to use the least square method to plot the experimental points, which are linearly correlative. The obtained curves have the smallest errors and higher accuracy.

If there are more than two variables, when fitting $y=f(x, z)$, fix a variable (for example, make z fixed) to obtain the relationship of $y \sim x$, thus to obtain a set of lines in the condition of different z.

3. Representation of experimental data with mathematical equation

When a group of experimental data are represented with tabulation and graphic methods, the relationships between parameters and variables are required to be further described with mathematical equations in certain situations. The representation of relationships between variables with mathematical formulas is simple and convenient. The method is described as follows: draw up the curves of experimental data and compare them with typical curves of known functions (linear equations, exponential equations, parabolic equations, circular and elliptic equations, etc). Select the proper function and calculate the constants and coefficients to obtain empirical formulas.

There are many empirical formulas which can be used to calculate constants and coefficients. Linear graphic method, average method and the least square method are most commonly used.

(1) Calculation of undetermined coefficients with linear graphic method

When the studied functional relationship is linear or linearized with straight line method, the coefficients can be represented by $y=a+bx$. The slope of the straight line $\left(\dfrac{\Delta y}{\Delta x}\right)$ is the value of b in the equation. The intercept of straight line on the y axis is the value of a in the equation.

It should be noted that the calculation of slope and intercept in the straight line equation on a logarithmic coordinate system is different from that on the rectangular coordinate system.

(2) Calculation of undetermined coefficients with average method

The ideal curve should be the curve which can make the algebraic sum of deviation of measured values to be zero. Assuming that the plotted ideal curve is a straight line and its equation is

$$y=a+bx \tag{2-39}$$

Assuming that the measured value is x_i, y_i, substitute x_i into equation (2-39), the value of y is y_i', then

$$y_i'=a+bx_i \tag{2-40}$$

Theoretically, $y_i'=y_i$. However, measured points generally deflect from the straight line due to errors in measurement, so $y_i' \neq y_i$. Assuming that the difference between y_i and y_i' is Δi, then

$$\Delta i = y_i - y_i' = y_i - (a + bx_i) \quad (2\text{-}41)$$

It is better to plot a straight line so that the sum of difference can be zero. Assuming the number of measured values is N

$$\sum \Delta i = \sum y_i - Na - b \sum x_i = 0 \quad (2\text{-}42)$$

Determine a, b according to equation (2-42), then the straight line taking a, b as constant is the desired ideal curve.

As equation (2-42) includes two unknown values a and b, the measured values are divided into equal or approximately equal sets in the order of experimental data. Establish their corresponding equations and combine both equations into simultaneous equation. a, b can be obtained by solving it.

(3) Calculation of undetermined coefficients with least square method

Deviation may be "positive" or "negative". In the processing of data, positive and negative deviation may be destructive, which cannot reveal the nature of deviation between numeric values. However, the square of deviation is positive. If the sum of squares of deviation is smallest, it indicates that the deviation is smallest. Least square method is defined as follows: the ideal curve is the curve which can minimize the sum of squares of the deviation between points and curves. It is derived based on the theory of errors. As can be known from equation (2-41)

$$\Delta i^2 = (y_i - y_i')^2 = [y_i - (a + bx_i)]^2 \quad (2\text{-}43)$$

The condition for obtaining the best value is

$$\sum \Delta i^2 = \sum [y_i - (a + bx_i)]^2 \rightarrow \text{minimum} \quad (2\text{-}44)$$

When the partial differential value of the logarithm a and b in equation (2-44) is 0, this condition can be met.

$$\frac{\partial (\sum \Delta i^2)}{\partial a} = -2 \sum [y_i - (a + bx_i)] = 0$$

So
$$\sum y_i = Na + b \sum x_i \quad (2\text{-}45)$$

$$\frac{\partial (\sum \Delta i^2)}{\partial b} = -2 \sum x_i [y_i - (a + bx_i)] = 0$$

So
$$\sum x_i y_i = a \sum x_i + b \sum x_i^2 \quad (2\text{-}46)$$

Equation (2-45) and equation (2-46) are the general formulas used in the calculation of constants a and b in the straight line equation obtained from least square method.

$$b = \frac{\begin{vmatrix} \sum y_i & N \\ \sum x_i y_i & \sum x_i \end{vmatrix}}{\begin{vmatrix} \sum x_i & N \\ \sum x_i^2 & \sum x_i \end{vmatrix}} = \frac{\sum x_i \sum y_i - N \sum x_i y_i}{(\sum x_i)^2 - N \sum x_i^2} \quad (2\text{-}47)$$

$$a = \frac{\begin{vmatrix} \sum x_i & \sum y_i \\ \sum x_i^2 & \sum x_i y_i \end{vmatrix}}{\begin{vmatrix} \sum x_i & N \\ \sum x_i^2 & \sum x_i \end{vmatrix}} = \frac{\sum x_i \sum x_i y_i - \sum y_i \sum x_i^2}{(\sum x_i)^2 - N \sum x_i^2} \tag{2-48}$$

Due to the smallest sum of square of deviation of the least square method, their random errors are always the smallest. However, the manual calculation is complicated with this method. There is no such disadvantage if it is calculated by a computer. When experimental data fit well with a straight line, the average method and graphic method can be used.

In addition to the graphic method, average method and least square method described above, constants in mathematical equations can be obtained by other methods such as selected point method and regression analysis method.

4. Correlation coefficient r and significance test

(1) Correlation coefficient r

The relationship between variables of experimental data is uncertain. Each value of one variable corresponds to the entire set values. When x is changed, the distribution of y is changed in some way. In this case, the relationship between the variable x and y is correlative.

In the calculation of the regression equation, it is unnecessary to assume that there must be a correlation between two variables. Even if there is a set of completely disordered points, the relationship between x and y can be represented by a straight line obtained from the least square method. However, it is pointless. Only when there is a linear relationship between two variables are they suitably performed with linear regression. Therefore, a quantitative index is required to describe the degree of linear relationship between two variables. Therefore, a term called correlation coefficient r, a statistical quantity, is introduced to determine the linear correlation between two variables.

$$r = \frac{\sum (x_i - \bar{x})(y_i - \bar{y})}{\sqrt{\sum (x_i - \bar{x})^2 \sum (y_i - \bar{y})^2}} \tag{2-49}$$

where
$$\bar{x} = \frac{\sum x_i}{n}$$

$$\bar{y} = \frac{\sum y_i}{n}$$

The theory of probability can prove that the absolute value of the correlation degree between any two random variables is not greater than 1, that is

$$|r| \leqslant 1 \text{ or } 0 \leqslant |r| \leqslant 1 \tag{2-50}$$

The physical significance of r: it indicates the degree of linear correlation between two random variables x and y. The following text will describe it with several instances.

① When $r = \pm 1$, experimental values (x_i, y_i) of n sets all fall on the straight line $y = a + bx$, they are completely correlative.

② When $0<|r|<1$, it indicates the most common case, there is a linear relationship between x and y. When $r>0$, $b>0$, the distribution of scattered points on the graph is y increases with the increase of x, there is a positive correlation between x and y. When $r<0$, $b<0$, y decreases with the increase of x, there is a negative correlation between x and y. The smaller $|r|$ is, the farther the scattered points are away from the regression line, the more the points are scattered. When $|r|$ is close to 1, that is, experimental values (x_i, y_i) of n sets are closer to $y=a+bx$, the relationship between the variable y and x is more linear.

③ When $r=0$, there is absolutely no linear relationship between the variables. It should be noted that the nonexistence of linear relationship does not mean that there are no other functional relations.

(2) Significance test

As mentioned above, the closer the absolute value of correlation coefficient r to 1, the larger the linearity of y and x is. But what is the closeness of $|r|$ to 1 can it suggest the linear correlation between x and y? It is necessary to conduct a significance test on correlation coefficients. Only when $|r|$ reaches a certain degree can the regression lines be used to approximately represent the relationship between x and y. At this point, the correlation is significant. Generally, the value of correlation coefficient r to make correlation significant is relevant to the number of experimental data points n. Therefore, only when $|r|>r_{\min}$ can the linear regression equations be used to describe the relationship between variables. The values of $r_{\min}$ can be found in Table 2-3. With the table, corresponding $r_{\min}$ can be found according to the number of data points n and significance level a.

Table 2-3 Significance test correlation coefficient

Degrees of freedom($n-2$)	Significance level 5%	Significance level 1%	Degrees of freedom($n-2$)	Significance level 5%	Significance level 1%	Degrees of freedom($n-2$)	Significance level 5%	Significance level 1%
1	0.997	1.000	16	0.468	0.590	35	0.325	0.418
2	0.950	0.990	17	0.456	0.575	40	0.304	0.393
3	0.878	0.959	18	0.444	0.561	45	0.288	0.372
4	0.811	0.917	19	0.433	0.549	50	0.273	0.354
5	0.754	0.874	20	0.423	0.537	60	0.250	0.325
6	0.707	0.834	21	0.413	0.526	70	0.232	0.302
7	0.666	0.798	22	0.404	0.515	80	0.217	0.283
8	0.632	0.765	23	0.396	0.505	90	0.205	0.267
9	0.602	0.735	24	0.388	0.496	100	0.195	0.254
10	0.576	0.708	25	0.381	0.487	125	0.174	0.228
11	0.553	0.684	26	0.374	0.478	150	0.159	0.208
12	0.532	0.661	27	0.367	0.470	200	0.138	0.181
13	0.514	0.641	28	0.361	0.463	300	0.113	0.148
14	0.497	0.623	29	0.355	0.456	400	0.098	0.128
15	0.482	0.606	30	0.349	0.449	1000	0.062	0.081

Generally $a=1\%$ or $a=5\%$ can be taken. If $|r|\geqslant 0.798$, it indicates that the linear correlation is significant at the level of $a=0.01$. When $0.798\geqslant|r|\geqslant 0.666$, it indicates that the linear correlation is significant at the level of $a=0.05$. When $|r|\leqslant 0.666$, it indicates that the correlation is not significant. At this point, there is no linear relationship between x and y. It is unnecessary to draw the regression line. The smaller the value of a is, the higher the significance is.

Chapter 3 Overview of Virtual Simulation Chemical Engineering Principle Experiment

I Software operating environment

Operating system: Windows XP/7/8
CPU: Intel Dual-core CPU @ 2.40GHz or above
Memory: More than 2G
Video card: Video memory of more than 1G, bit width of more than 256bit

II Virtual simulation system of chemical engineering principle experiment startup

Double click the installation package on a computer in normal operation. The virtual simulation software can be installed.

Taking the experiment of packed column absorption for example:

The start screen of absorption simulation experimental system, Windows system is shown in Figure 3-1.

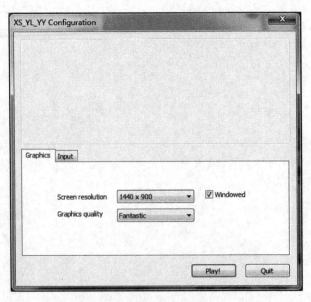

Figure 3-1 Start screen of absorption simulation

Select the desired screen resolution.

Select the desired graphics quality.

III Menu function of virtual simulation chemical engineering principle experiment

When the mouse cursor moves to the upper area of the screen, the hidden menu automatically pops up as shown in Figure. 3-2.

Figure 3-2 PC hidden menu

1. Task

Basic Operation, Security, Introduction, Structure and Case are included in the task menu. Click the menu then the level window pops up as shown in Figure 3-3.

① Basic Operation

The students who have never used the virtual simulation software should read the guidance to understand the basic operations of the software.

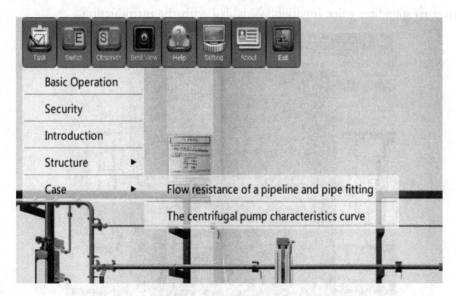

Figure 3-3 PC level menu

② Security

Click the button and corresponding interface for introduction pops up as shown in Figure 3-4. Then corresponding voice function is enabled with introduction.

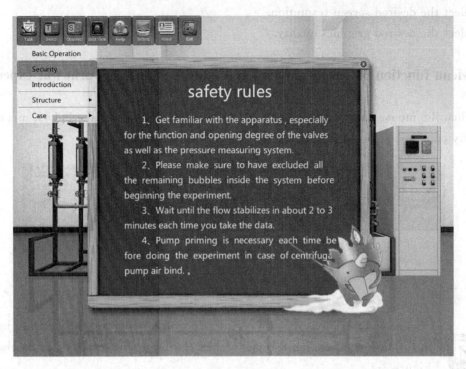

Figure 3-4　Interface of Security

③ Introduction

Click the button of "Introduction", corresponding interface pops up as shown in Figure 3-5. Then corresponding voice function is enabled with the introduction.

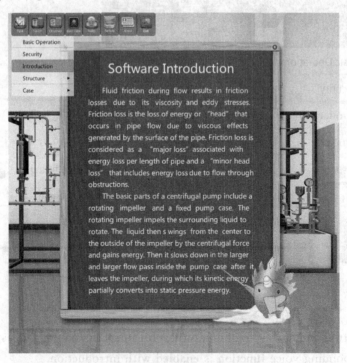

Figure 3-5　Interface of Introduction

④ Structure

After clicking the button "Introduction" of comprehensive fluid mechanic experiment, students can have a whole understanding of the comprehensive fluid mechanic experiment. Figure 3-6 shows the diagram of the structure.

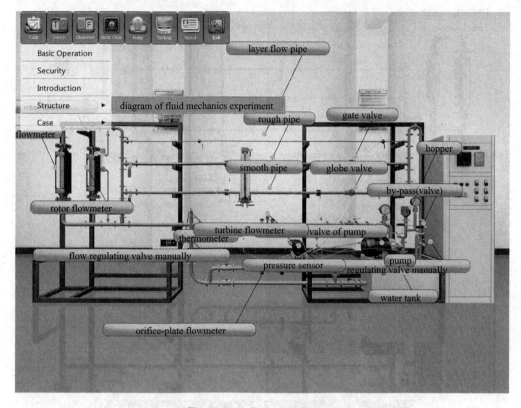

Figure 3-6   Structure diagram

⑤ Case

Click the interface of "Case" and the interface pops up as shown in Figure 3-7.

After an appropriate case is selected, there is a brief description of the experiment as shown in Figure 3-8.

After closing the case description, students can enter the interface of teaching and practice as shown in Figure 3-9. Different interfaces with corresponding functions will pop up at different stages of learning. Users can learn relevant operations according to the text and voice prompts.

In the learning process, all interactive functions are temporarily blocked. The system will automatically perform the operations of case teaching accompanied by text and voice prompts. Users can learn the operation procedures of experiment.

In the process of practice, the system will automatically enable corresponding functions according to current procedures to guide users to learn the operation procedures systematically.

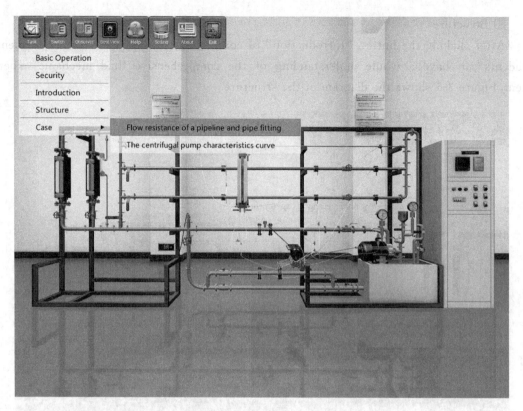

Figure 3-7 Interface of Case

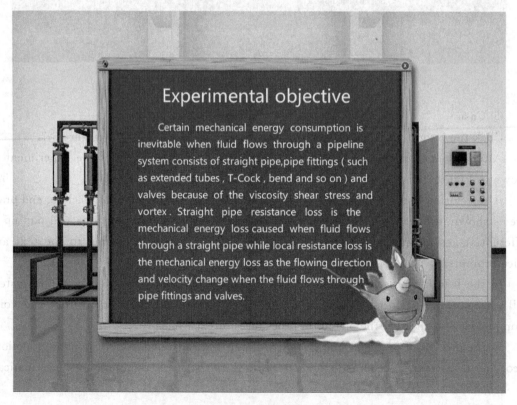

Figure 3-8 Interface of Case description

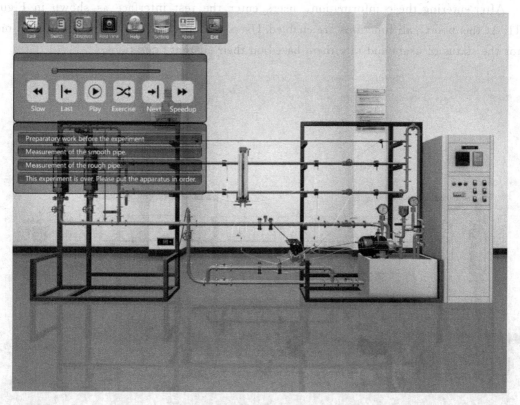

Figure 3-9 Interface of Case teaching and practice

2. Practice and test

Switch

When users are familiar with the operating procedures, they can press the button of "Switch" in the hidden menu to enter the test mode for corresponding examination. When the program is started for test, the system pops up an input window which requires users to enter their user name and password as shown in Figure 3-10.

Figure 3-10 Input window of test information

Chapter 3 Overview of Virtual Simulation Chemical Engineering Principle Experiment

After entering these information, users enter the test interface as shown in Figure 3-11. At this point, all functions are enabled. Users can operate freely. The system will monitor the status of users and rate them based on their current operations.

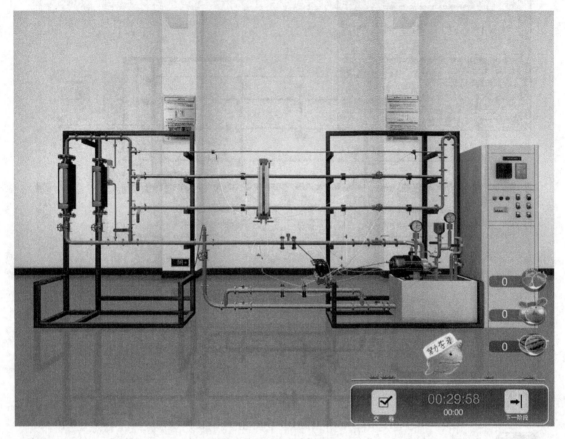

Figure 3-11 Test interface

When users cannot complete the current step, they can click on the "Next Stage" button to move on to the next examination but they will automatically discard the current score. When users click on the "Finish" button, the system will pop up a confirmation menu. After confirming, the software will automatically send the results to the system desktop. Users can check the results by themselves.

3. Observer

This function is enabled when users perform practice and test (press the button if there are no corresponding actions when entering the scene). The mode consists of "Focus mode" and "Scene mode". Users can view the lab in this mode.

In the scene mode, camera lens can be changed with keyboard: W, A, S and D mean move around, Q and E mean rotate left and rotate right, Z and X mean adjust up and adjust down, the space bar means skip.

Scene mode	Front	Rear	Left	Right	Up	Down
	W	S	A	D	Z	X

Observer

In the focus mode, camera lens can be changed with mouse: press the middle mouse to rotate the scene. Press the middle key and the right key at the same time to shift the scene.

Focus mode	Front and rear	Camera direction	Position movement
	Roller rolling	Click the roller, move the mouse	Click the roller and right key, move the mouse

4. Best view

Best view

In the process of practice and test, users can have the best position for viewing by pressing the button.

5. Help

Help

In the process of practice, users can click on the button to highlight the object to be operated.

6. Part prompt

Part prompt

In the mode of part prompt, there will be a corresponding name displayed when moving the mouse over the object. If pressing the button, the menu will become gray and there will be no prompts when moving the mouse over the object.

7. About us

About

Click the button and a general information on the development of company will display.

8. Exit

Exit

When clicking the button, an interface of exit pops up.

Chapter 4　Comprehensive Fluid Mechanic Experiment—Determination of Flow Resistance in Pipe

I　Experimental purposes and requirements

1. To grasp the general method to measure the resistance loss of fluid when flowing through straight pipe and pipe fittings (valve).
2. To determine the relationship between straight pipe friction coefficient λ and Reynolds number Re and verify the relationship between λ and Re in general turbulent region.
3. To measure the local resistance coefficient ξ of fluid when flowing through pipe fittings (valve).
4. To identify pipe fittings and valves of pipeline and understand their functions.

II　Experiment principle

Some mechanical energy would inevitably be consumed due to viscous shear stresses and eddy currents of fluid when it flows in pipe. The losses include the resistance loss of straight pipe due to the flowing of fluid through the straight pipe and the local resistance loss caused by the changes in flow direction or pipe size and shape.

1. Flowmeter checking

Check readings of flowmeter through specified interval times.

2. Reynolds number Re

$$Re = \frac{du\rho}{\mu} \tag{4-1}$$

$$u = \frac{V}{900\pi d^2} \tag{4-2}$$

where　Re—Reynolds number, dimensionless;
　　　d—inner diameter of straight pipe, m;
　　　u—average velocity of fluid flowing in pipe, m/s;
　　　ρ—fluid density, kg/m³;
　　　μ—fluid viscosity, kg/(m·s);

V—fluid flow measured with turbine flowmeter, m³/h.

3. Measurement of friction coefficient λ of straight pipe resistance

When fluid steadily flows in a horizontal straight pipe with the same diameter, the resistance loss is

$$h_f = \frac{\Delta p_f}{\rho} = \frac{p_1 - p_2}{\rho} = \lambda \frac{l}{d} \times \frac{u^2}{2} \tag{4-3}$$

That is

$$\lambda = \frac{2d \Delta p_f}{\rho l u^2} \tag{4-4}$$

where λ—friction coefficient of straight pipe resistance, dimensionless;

Δp_f—pressure drop when fluid flows through the l meter straight pipe, Pa, measured with differential pressure sensor;

h_f—loss of mechanical energy of fluid per unit mass flowing through the straight pipe of l meter, J/kg;

l—straight pipe length, m.

4. Measurement of local resistance coefficient ξ

Method of resistance coefficient: the loss of mechanical energy of a fluid when it flows through a tube (valve) can be represented by certain times of average kinetic energy of the fluid when it flows in a small diameter tube. That is

$$h'_f = \frac{\Delta p'_f}{\rho} = \xi \frac{u^2}{2} \tag{4-5}$$

Therefore

$$\xi = \frac{2 \Delta p'_f}{\rho u^2} \tag{4-6}$$

where h'_f—loss of mechanical energy of fluid per unit mass when flowing through a certain pipe (valve), J/kg;

ξ—local resistance coefficient, dimensionless;

$\Delta p'_f$—local resistance pressure drop, Pa.

Measurement of local resistance pressure drop: the flow of fluid can be affected by the pipe fittings, therefore the actual pressure drop of pipe fittings is measured by subtracting the straight pipe pressure drop from the total pressure drop (total length l'). The straight pipe pressure drop can be obtained from the experimental results of resistance Δp_f in the straight pipe (length l).

$$\Delta p'_f = \sum \Delta p - \frac{l'}{l} \Delta p_f \tag{4-7}$$

where $\sum \Delta p$—pressure drop of the pipe fitting (valve) and straight pipe (length l'), Pa, measured with differential pressure sensor;

l'—length of straight pipe at $\sum \Delta p$ measuring point (both sides of valve), m.

III Experimental apparatus and process

1. Experimental apparatus

The experimental apparatus (as shown in Figure 4-1) includes a water storage tank, a centrifugal pump, pipes of different diameters and materials, valves, pipe fittings, flowmeters and differential pressure sensors. The pipelines are composed of three long straight pipes in parallel, respectively used for determining the laminar resistance of straight pipe, the resistance coefficient of rough straight pipe and the resistance coefficient of smooth straight pipe from top to bottom. Meanwhile, the rough straight pipe and the smooth straight pipe are equipped with a gate valve and a globe valve which are used for determining the local resistance coefficient of different valves. The virtual simulation diagram of this experiment is shown in Figure 4-2.

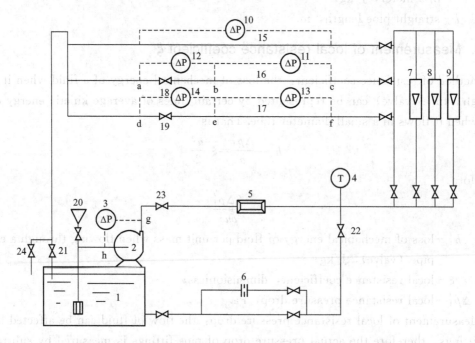

Figure 4-1 Comprehensive fluid mechanic experiment apparatus

1—Water storage tank; 2—Centrifugal pump; 3,10~14—Differential pressure sensor; 4—Thermometer; 5—Turbine flowmeter; 6—Orifice (or Venturi) flowmeter; 7~9—Rotameter; 15—Experimental section of laminar resistance; 16—Experimental section of rough pipe; 17—Experimental section of smooth pipe; 18—Gate valve; 19—Globe valve; 20—Intake funnel; 21,22—Flow control valve; 23—Pump outlet valve; 24—By-pass valve; a~h—Pressure measuring point

2. Parameter

The specifications of the apparatus are shown in Table 4-1.

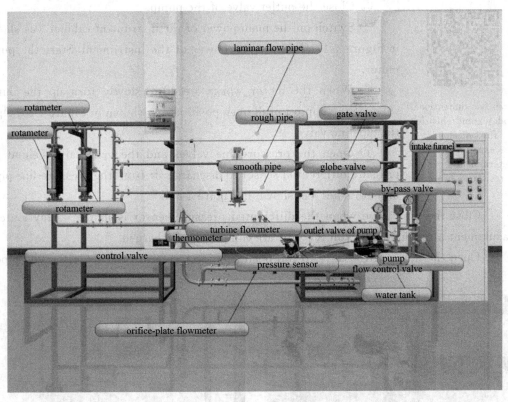

Figure 4-2 Diagram of virtual simulation of comprehensive fluid mechanic experiment

Table 4-1 Apparatus specification

Name	Type	Specification of straight pipe	Inner diameter /mm	Length /mm
Smooth pipe	Stainless steel pipe	Smooth straight pipe	21	ef=1000
Rough pipe	Galvanized iron pipe	Rough straight pipe	22	bc=1000
Local resistance	Globe valve	Straight pipe of both sides of globe valve(smooth pipe)	21	de=660
Local resistance	Gate valve	Straight pipe of both sides of gate valve(rough pipe)	22	ab=680

Ⅳ Operation steps of virtual simulation experiment

1. Experiment preparation

① Open the valve of the intake funnel.

② Use a beaker to fill the intake funnel of the pump with water (as shown in Figure 4-3). Close the valve of the intake funnel.

Scan two-dimensional code: comprehensive fluid mechanic experiment—Determination of flow resistance in pipe

③ Close the outlet valve of the pump.

④ Switch on the main power of the instrument cabinet (as shown in Figure 4-4). Turn on the power of the instrument. Start the pump motor.

⑤ When the motor works steadily, slowly turn up the outlet valve of the pump to full open position (as shown in Figure 4-5). Open the pipeline valve.

⑥ Open the pressure-line valves and the vent valve behind the pressure differential sensor to discharge air from the pressure-line pipes on both sides (as shown in Figure 4-6).

⑦ Close the vent valve when the differential pressure sensor is in the steady state of measurement after the bubbles in the pipe are completely discharged.

⑧ Open the flow control valve to full open position to remove the air in the test pipe.

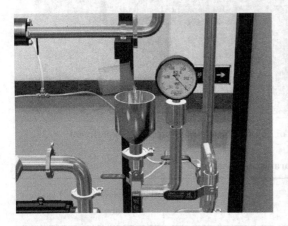

Figure 4-3 Infusion of centrifugal pump

Figure 4-4 Operation of control cabinet

Figure 4-5 Open the pump outlet valve

Figure 4-6 Open the vent valve behind the pressure difference sensor

2. Measurement of smooth pipe and pipe fitting (globe valve)

① Close the pressure-line valve and the main valve at the test section of the rough pipe.

② Open the pressure-line valve and the main valve at the test section of the smooth pipe, measure the smooth pipe.

③ Confirm that the globe valve of the pipe and the flow control valve are completely opened.

④ When the flow data on the display are stable, record the maximum flow rate in the pipeline.

⑤ Distribute the flow rate according to the maximum flow rate in the pipe. Adjust the flow control valve (as shown in Figure 4-7).

⑥ After the flow rate is changed and stable, record the data (as shown in Figure 4-8).

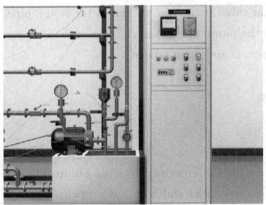

Figure 4-7 Open the flow control valve

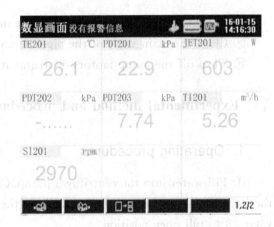

Figure 4-8 Data display

3. Measurement of rough pipe and pipe fitting (gate valve)

① Close the pressure-line valve and the main valve at the test section of the smooth pipe.

② Open the main valve (as shown in Figure 4-9) and the pressure-line valve of the rough pipe to measure the test section.

③ Confirm that the gate valve of the pipe is completely opened (as shown in Figure 4-10). Open the flow control valve to full open position.

④ When the flow data on the display are stable, record the maximum flow rate in the pipeline.

⑤ Distribute the flow rate according to the maximum flow rate in the pipe. Adjust the flow control valve (as shown in Figure 4-7).

⑥ After the flow rate is changed and stable, record the data (as shown in Figure 4-8).

4. Equipment arrangement after experiment

① Close the flow control valve.

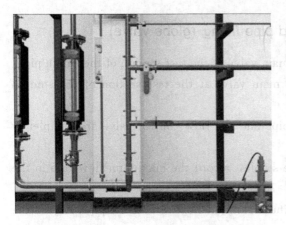

Figure 4-9 Open the valve of the rough pipe

Figure 4-10 Confirm that the gate valve of the pipe is completely opened

② Close the pressure-line valve and the main valve at the test section of the rough pipe.
③ Close the main valve of the pipeline and the pump outlet valve.
④ Turn off the pump motor, instrument power and main power.

Ⅴ Experimental method and procedure

1. Operating procedures

① Fill water into the centrifugal pump. Close the outlet valve 23 of the pump. Turn on the power to start the pump motor. When the motor works stably, slowly open the outlet valve 23 to full open position.

② Open the vent valve behind the differential pressure sensor, and discharge air from the pressure tube. Then close the vent valve to keep the differential pressure sensor in the state for measurement.

③ Open the bypass valve 24. Calibrate the flowmeters using several selected flow rate.

④ Design the experimental record sheet, select test tube, open the flow control valve 21 and rationally distribute the flow rate. Record the data of differential pressure, flow rate and temperature after the flow rate is changed and becomes stable every time.

⑤ At the end of the experiment, close the outlet valve 23. Turn off the pump motor and clean the experiment apparatus.

2. Attention

① Be familiar with the experimental apparatus, particularly the functions of valves and pressure measuring points.

② Be sure to completely remove the bubbles in the system prior to conduct the experiment.

③ Be sure to record the readings after the flow is stable (for about 2~3min) in each measurement.

VI Experimental report

1. Calculate Reynolds number Re, resistance coefficient λ of smooth and rough pipes and local resistance ξ of pipe fittings (valve) by using the experimental data.

2. Based on the experimental results of smooth and rough pipes, fit out the curves between λ and Re on the double-logarithm coordinate system and estimate the relative and absolute roughness of the pipe comparing with Moody graph in relevant manuals.

3. According to the experimental results of local resistance, analyze the change of the local resistance (valve) coefficient ξ with Reynolds number Re and compare it with experience value in relevant manuals.

4. Analyze and discuss the rationality of the above experimental results.

VII Questions

1. How to ensure that air is completely removed from the pipe?

2. Is the relationship between λ and Re measured with the medium of water applied to other fluids? How is it applied?

3. Can the data of $\lambda \sim Re$ measured on different apparatuses (including pipes of different diameters) under different water temperatures be correlated on the same curve?

4. What is the impact of burrs at the edge of the pressure port and hole or non-vertical installation on the measurement of static pressure?

5. Which differential gauges can be shared in differential pressure test? Why?

Chapter 5 Comprehensive Fluid Mechanic Experiment—Determination of Centrifugal Pump Characteristic Curve

I Experimental purposes and requirements

1. To understand the structure and characteristics of centrifugal pump and be familiar with the use of centrifugal pump.

2. To measure the characteristics of centrifugal pump in the operation at constant rotate speed and draw the characteristic curves.

3. To understand the working principles and usage of instruments include differential pressure transmitter and turbine flowmeter.

II Experiment principle

For a fixed rotational speed n, the relationship between actual head H, power consumption N, pump efficiency η and pumping flow rate Q can be expressed as a curve which is called centrifugal pump characteristic curve. The curve is the basis for selecting and sizing centrifugal pump, which reflects the operational performance of the pump.

Centrifugal pump characteristic curve can be expressed by the three functions below

$$H = f_1(Q); \quad N = f_2(Q); \quad \eta = f_3(Q)$$

These functional relationships can be tested by experiment, the methods are as follows.

1. Flow rate Q

Fluid flow rate Q can be measured by flowmeter.

2. The head of pump H

Take the sections 1 and 2 at the place where the vacuum gauge and pressure gauge are located at the inlet and the outlet of pump to derive the equation of mechanical energy balance.

$$z_1 + \frac{p_1}{\rho g} + \frac{u_1^2}{2g} + H = z_2 + \frac{p_2}{\rho g} + \frac{u_2^2}{2g} \tag{5-1}$$

Because the inlet and outlet diameter of the pump used in this experiment is the same, there is no big difference between flow velocities and the difference of squares of velocity can be ignored, then

$$H = (z_2 - z_1) + \frac{p_2 - p_1}{\rho g} = H_0 + \frac{\Delta p}{\rho g} \tag{5-2}$$

$$H_0 = (z_2 - z_1)$$

$$\Delta p = (p_2 - p_1)$$

where H_0—the potential difference between the pressure taps at the inlet and outlet of pump, the potential difference is 0.1m in this experiment;

ρ —fluid density, kg/m^3;

g —gravitational acceleration, m/s^2;

p_1, p_2 —the gauge pressure of pump inlet and outlet, Pa;

u_1, u_2 —the flow velocity at the inlet and outlet of pump, m/s;

z_1, z_2 —the installation height of vacuum gauge and pressure gauge, m;

Δp —differential pressure between the pump outlet and inlet, Pa, measured with a differential pressure sensor.

3. Shaft power of pump N

Shaft power N is the input work to centrifugal pump from electrical machine per unit time

$$N = N_p k_e k_t \tag{5-3}$$

where N_p—indication of electric power meter, W;

k_e —motor efficency, lab offer;

k_t —transmission efficiency, $k_t = 1$ as motor is connected with pump.

4. Efficiency η

The power losses in pump are produced by hydraulic loss, mechanical loss and volumetric loss. Effective power N_e is the actual power gained by fluid when it flows through the pump per unit time; the efficiency η of pump is the ratio between effective power N_e and shaft power N.

The effective power N_e of pump

$$N_e = HQ\rho g \tag{5-4}$$

Pump efficiency η

$$\eta = \frac{N_e}{N} \times 100\% \tag{5-5}$$

5. Calibration on the basis of speed

The characteristic curves of pump are obtained in the experiment at constant rotate speed. However, the speed of induction motor changes with its torque, thus the rotate speed n at multiple experimental points would change with flow rate Q. Therefore, before the characteristic curves are fitted, the measured data are required to be converted to the data at a certain constant speed n' (it can be the rated speed of centrifugal pump).

Flow rate

$$Q' = Q\frac{n'}{n} \tag{5-6}$$

Head

$$H' = H\left(\frac{n'}{n}\right)^2 \tag{5-7}$$

Shaft power

$$N' = N\left(\frac{n'}{n}\right)^3 \tag{5-8}$$

Efficiency

$$\eta' = \frac{Q'H'\rho g}{N'} = \frac{QH\rho g}{N} = \eta \tag{5-9}$$

Ⅲ Experimental apparatus and process

Experimental apparatus includes water storage tank, centrifugal pump, turbine flowmeter and differential pressure sensors.

Water flow rate is measured with a turbine flowmeter. Pressure difference between the pump inlet and outlet is measured with a differential pressure sensor. Shaft power is measured with a power meter. Fluid temperature is measured with a Pt100 temperature sensor.

The experimental apparatus is shown in Figure 4-1 and Figure 4-2.

Ⅳ Operation steps of virtual simulation experiment

1. Experiment preparation

① Open the valve of the intake funnel.

② Use a beaker to fill the intake funnel of the pump with water (as shown in Figure 4-3). Close the valve of the intake funnel.

③ Close the outlet valve of the pump.

④ Switch on the main power of the instrument cabinet (as shown in Figure 4-4). Turn on the power of the instrument.

⑤ Start the pump motor.

⑥ When the motor works steadily, slowly turn up the outlet valve of the pump to full open position (as shown in Figure 4-5).

⑦ Open the pressure-line valve at differential pressure sensor (as shown in Figure 5-1).

⑧ Open the vent valve behind the differential pressure sensor to discharge air from the pressure-line pipe.

⑨ Close the vent valve when the differential pressure sensor is in the steady state for measurement after the bubbles in it are completely discharged.

Scan two-dimensional code: comprehensive fluid mechanic experiment—Determination of centrifugal pump characteristic curve

⑩ Fully open the control valve to discharge the air in the test pipe (as shown in Figure 5-2).

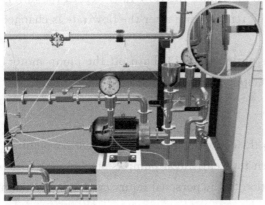
Figure 5-1 Open the pressure-line valve

Figure 5-2 Open the flow control valve

2. Experimental measurement

① When the flow data on the screen are stable, record the maximum flow rate in the pipeline.

② Record the differential pressure, flow rate, motor power and fluid temperature under the conditions of different flow.

③ Distribute the flow rate according to the maximum flow rate in the pipe. Adjust the flow control valve.

④ After the flow rate is changed and stable, record the flow rate.

⑤ Record the related data of the apparatus, such as the type and rated flow rate of the centrifugal pump.

3. Equipment arrangement after experiment

① Close the flow control valve.
② Close the pressure-line valves and the test pipe valves.
③ Close the outlet valve of the pump.
④ Turn off the pump motor, the power of the instrument and main power.

V Experimental method and procedure

1. Operating procedures

① Fill water into the centrifugal pump. Close the outlet valve 23 of the pump. Turn on the main power on the instrument cabinet and instrument power to start the water pump. When the pump motor works stably, slowly open the outlet valve 23 to full open position.

② Discharge the air from the differential pressure sensor. Then close the vent valve to keep the differential pressure sensor in the state for measurement.

③ Slowly open the flow control valve 22 and distribute the flow rate. Record the differential pressure, flow rate, motor power and fluid temperature after the flow rate is changed and stable every time.

④ At the end of the experiment, close the outlet valve 23. Turn off the pump motor, the instrument power and main power. Reset the apparatus.

2. Attention

① Prior to each experiment, fill water into the centrifugal pump to prevent air binding.

② Make sure the outlet valve is closed when the centrifugal pump is started.

③ After the pump starts, please pay attention to the personal injury caused by the high rotation speed of the pump.

Ⅵ Experimental report

1. Fit the characteristic curves ($H\sim Q$, $N\sim Q$, $\eta\sim Q$) of the pump at a certain rotate speed in the same coordinate system.

2. Analyze the results to determine the most appropriate operating range of the pump.

Ⅶ Questions

1. Why must the outlet valve of a centrifugal pump be closed when the pump is started, analyze based on the measured data.

2. Why does water need to be filled into the centrifugal pump before the pump is started? If the centrifugal pump cannot work normally, what are the possible causes?

3. Experiment data show that the larger of the water flow rate, the larger of the vacuum degree in pump entrance, why?

4. Do the readings of the pressure gauge gradually increase if the outlet valve is not opened after the pump is started? Why?

5. Is it reasonable to install a valve on the inlet pipe of a centrifugal pump which works normally? Why?

Chapter 6 Determination Experiment of Convective Heat Transfer Coefficient

I Experimental purposes and requirements

1. To grasp the method to measure the convective heat transfer coefficient of air in the heat transfer pipe and understand the factors affecting heat transfer coefficient and the approach to enhance heat transfer.
2. To deduce the measured data into dimensionless equation in the form of $Nu = ARe^n$ and compare it with the recognized equations in textbooks.
3. To understand the principles and methods of automatic control of temperature, heating power and air flow rate.

II Experiment principle

In the process of industrial production, the wall-type heat exchange is often adopted. Wall-type heat exchange means that heat is exchanged through solid walls (heat transfer component) on both sides of which hot and cold fluids flow without direct contact.

The apparatus in this experiment focuses on steam-air heat exchange, including regular tube and enhanced tube. Heat is indirectly exchanged between steam and air through a copper tube. Air flows in the copper tube and steam flows outside, performing counterflow heat exchange. As for enhanced copper tube, spring is installed to increase the absolute roughness, thus increases the turbulence of air flow and results in effective heat exchange.

1. Basic principle of wall-type heat transfer

As shown in Figure 6-1, wall-type heat transfer consists of convective heat transfer from hot fluid to solid wall, solid wall heat conduction and convective heat transfer from solid wall to cold fluid.

When the wall-type heat transfer components work stably in the process of heat transfer

$$Q = m_1 c_{p1}(T_1 - T_2) = m_2 c_{p2}(t_2 - t_1)$$
$$= \alpha_1 A_1 (T - T_W)_m = \alpha_2 A_2 (t_W - t)_m$$
$$= KA \Delta t_m \qquad (6-1)$$

The logarithmic mean temperature difference between hot fluid and solid wall can be calculated from equation (6-2).

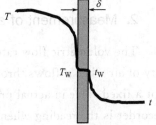

Figure 6-1 Wall-type heat transfer

$$(T-T_W)_m = \frac{(T_1-T_{W1})-(T_2-T_{W2})}{\ln\dfrac{T_1-T_{W1}}{T_2-T_{W2}}} \qquad (6\text{-}2)$$

The logarithmic mean temperature difference between solid surface and cold fluid can be calculated from equation (6-3).

$$(t_W-t)_m = \frac{(t_{W1}-t_1)-(t_{W2}-t_2)}{\ln\dfrac{t_{W1}-t_1}{t_{W2}-t_2}} \qquad (6\text{-}3)$$

The logarithmic mean temperature difference between hot and cold fluids can be calculated from equation (6-4).

$$\Delta t_m = \frac{(T_1-t_2)-(T_2-t_1)}{\ln\dfrac{T_1-t_2}{T_2-t_1}} \qquad (6\text{-}4)$$

where
- Q —heat transfer capacity, W;
- m_1, m_2 —mass flow rate of hot and cold fluids, kg/s;
- c_{p1}, c_{p2} —specific heat of hot and cold fluids under qualitative temperature, J/(kg·℃);
- T_1, T_2 —temperature at the inlet and outlet of hot fluid, ℃;
- t_1, t_2 —temperature at the inlet and outlet of cold fluid, ℃;
- T_{W1}, T_{W2} —wall temperature at the inlet and outlet of hot fluid, ℃;
- t_{W1}, t_{W2} —wall temperature at the inlet and outlet of cold fluid, ℃;
- α_1, α_2 —convective heat transfer coefficient of hot and cold fluids corresponding to solid wall, W/(m²·℃);
- A_1, A_2 —heat transfer area of hot and cold fluids, m²;
- $(T-T_W)_m$, $(t-t_W)_m$ —logarithmic mean temperature difference between hot and cold fluids and solid wall, ℃;
- K —total heat transfer coefficient based on heat transfer area A, W/(m²·℃);
- A —mean heat transfer area, m²;
- Δt_m —logarithmic mean temperature difference of cold and hot fluids, ℃.

The flow diagram of the experimental apparatus shows that the hot fluid is steam and the cold fluid is air for heat exchange in the experiment.

2. Measurement of air flow rate

The volumetric flow rate of air displayed on the paperless recorder is relevant to the density of air when it flows through the orifice plate. Considering that the inlet air temperature is not a fixed value in actual process, the volumetric flow rate of air displayed on the paperless recorder is the reading when the air density ρ_0 at the orifice plate is regarded as 1kg/m³ for easy processing. Therefore, if the actual density of air is not equal to the value, the actual volumetric flow rate of air should be corrected according to equation (6-5).

$$V' = \frac{V}{\sqrt{\rho_0}} \qquad (6\text{-}5)$$

Air mass flow rate m can be calculated from equation (6-6).

$$m = V'\rho_0 \qquad (6\text{-}6)$$

where V' —actual volumetric flow rate of air, m^3/s;

V —volumetric flow rate of air displayed on the paperless recorder, m^3/s;

ρ_0 —density of air in the orifice flowmeter, kg/m^3, in this experiment, ρ_0 is the density of air under the inlet temperature of t_1.

3. Measurement of convective heat transfer coefficient α of air in the double-pipe heat exchanger

(1) Newton cooling law

In the tube heater, water vapor passes within the annulus, air flows through the copper tube, water vapor condenses on the surface of copper tube and air is heated. After the heat transfer is stable, heat transfer at air side is

$$Q = m_2 c_{p2}(t_2 - t_1) = \alpha_2 A_2 (t_W - t)_m \qquad (6\text{-}7)$$

That is

$$\alpha_2 = \frac{m_2 c_{p2}(t_2 - t_1)}{A(t_W - t)_m} \qquad (6\text{-}8)$$

t_{W1} and t_{W2} are respectively the temperature of the inner wall of the heat exchange tube at the air inlet and outlet. When the tube material has high thermal conductivity, that is, the value of λ is large and the thickness of the wall is small, $T_{W1} \approx t_{W1}$ and $T_{W2} \approx t_{W2}$. T_{W1} and T_{W2} should respectively be the temperature (T3 and T6) of the outer wall of the heat exchange tube at the air inlet and the temperature (T2 and T5) of the outer wall of the heat exchange tube at the air outlet. Please see the flow diagrams in Figure 6-3 (horizontal tube) and Figure 6-4 (vertical tube).

Generally, it is difficult to directly measure the temperature of a solid wall, especially the temperature of the inner wall of tube. Therefore, the convective coefficient of heat transfer between fluid and solid wall is usually indirectly deduced by measuring the easily-measured temperature of fluid in engineering.

The following text will introduce two experimental methods [(2) and (3)] to measure the convective heat transfer coefficient α.

(2) Approximation method

The relationship between total heat transfer coefficient and convective heat transfer coefficient based on the inner wall area is

$$\frac{1}{K} = \frac{1}{\alpha} + R_{S2} + \frac{b d_2}{\lambda_1 d_m} + R_{S1}\frac{d_2}{d_1} + \frac{d_2}{\alpha_1 d_1} \qquad (6\text{-}9)$$

where d_1, d_2 —external and internal diameter of heat exchange tube, m;

d_m —logarithmic mean diameter of heat exchange tube for convection, m;

b —wall thickness of heat exchange tube, m;

λ_1—thermal conductivity of heat exchange tube, W/(m·℃);

R_{S1}, R_{S2}—fouling resistance at the outer and inner side of the heat exchange tube, m²·℃/W.

The total heat transfer coefficient K is calculated from equation (6-1).

$$K = \frac{Q}{A\Delta t_m} = \frac{m_2 c_{p2}(t_2 - t_1)}{A\Delta t_m} \tag{6-10}$$

When the experiment is carried out with this apparatus, the heat transfer coefficient α of convection between the air in the tube and tube wall is about dozens or hundreds of W/(m²·℃); steam condensation occurs outside the tube with a heat transfer coefficient α_1 of around 10^4 W/(m²·℃), so the thermal resistance $\frac{d_2}{\alpha_1 d_1}$ of condensation heat transfer can be ignored. Meanwhile, condensed steam is clean, so the fouling resistance $R_{S1}\frac{d_2}{d_1}$ at the outer side of the heat exchange tube can also be ignored. The heat transfer components in the experiment are made of copper with the thermal conductivity λ_1 of 383.8W/(m²·℃) and wall thickness of 1.5mm, so the thermal conduction resistance $\frac{bd_2}{\lambda_1 d_m}$ of the heat exchange tube can be ignored. If the fouling resistance R_{S2} at the inner side of heat exchange tube is also ignored, the following equation can be obtained from equation (6-9)

$$\alpha \approx K \tag{6-11}$$

Thus it can be seen that the smaller the ratio between the ignored heat transfer resistance and the heat transfer resistance of convection at the side of cold fluid is, the higher the accuracy of α measured with this method.

(3) Wilson diagram method

Heat is exchanged between air and steam in the double-tube heat exchanger. Air is heated by the steam in the annular space of the tube. When the air in the tube is fully developed into turbulence, the forced convective heat transfer coefficient at the air side can be expressed as

$$\alpha = Cu^{0.8} \tag{6-12}$$

By substituting equation (6-12) into equation (6-9), the result is expressed as follows

$$\frac{1}{K} = \frac{1}{Cu^{0.8}} + R_{S2} + \frac{bd_2}{\lambda_1 d_m} + R_{S1}\frac{d_2}{d_1} + \frac{d_2}{\alpha_1 d_1} \tag{6-13}$$

Based on the analysis of (2), the rear four terms at the right of equation (6-13) can be considered as constants in this experiment. The following equation can be obtained from equation (6-13).

$$\frac{1}{K} = \frac{1}{Cu^{0.8}} + \text{constant} \tag{6-14}$$

Equation (6-14) is the linear equation of $Y = kX + B$. The fitting based on $Y = \frac{1}{K}$ and $X = \frac{1}{u^{0.8}}$ is shown in Figure 6-2.

The equation can be obtained according to the slope $k=\tan\theta$ of experimental line

$$\alpha = Cu^{0.8} = \frac{u^{0.8}}{k} \quad (6\text{-}15)$$

4. Fitting of dimensionless number equation Nu~Re

After calculating the related dimensionless number Nu and Re based on the experimental data, fit the $Nu \sim Re$ line on the double-logarithm coordinate system to determine the fitting equation thus to obtain the experimental relation.

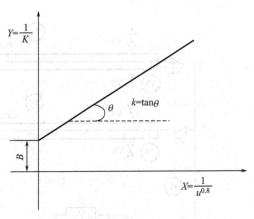

Figure 6-2 Linear fitting with wilson diagram method

$$Nu = ARe^n \quad (6\text{-}16)$$

where Nu —Nusselt number, $Nu = \frac{\alpha d}{\lambda}$, dimensionless;

Re —Reynolds number, $Re = \frac{du\rho}{\mu}$, dimensionless.

5. Empirical correlation between heat transfer and dimensionless number

For fluid force turbulent convection heat transfer in the round straight pipe, the empirical correlation

$$Nu = 0.023Re^{0.8}Pr^m \quad (6\text{-}17)$$

where Pr —Prandtl number, $Pr = \frac{c_p \mu}{\lambda}$, dimensionless;

λ —thermal conductivity of air under qualitative temperature, W/(m·℃);

μ —viscosity of air under qualitative temperature, Pa·s.

$m = 0.4$ when the fluid is heated, $m = 0.3$ when the fluid is cooled.

Under the conditions of the experiment, the change in Pr is small, so it can be considered as a constant, equation (6-17) can be converted into

$$Nu = 0.02Re^{0.8} \quad (6\text{-}18)$$

Ⅲ Experimental apparatus and process

The flow diagrams of the experiment are shown below. Figure 6-3 is horizontal pipe experimental facility, Figure 6-4 is vertical pipe experimental facility and Figure 6-5 is the virtual simulation diagram of horizontal pipe experimental apparatus. Each experimental apparatus is composed of a steam generator (transmitter), a orifice flowmeter, a frequency converter, a double-tube heat exchanger (enhanced tube and regular tube), several temperature sensors and intelligent display instruments.

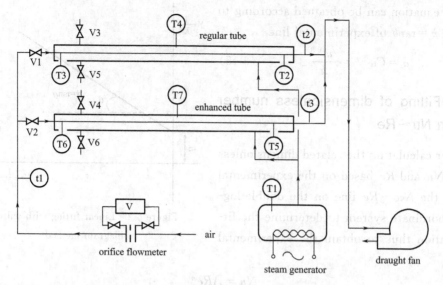

Figure 6-3　Measurement experiment of convective heat transfer coefficient of horizontal tube

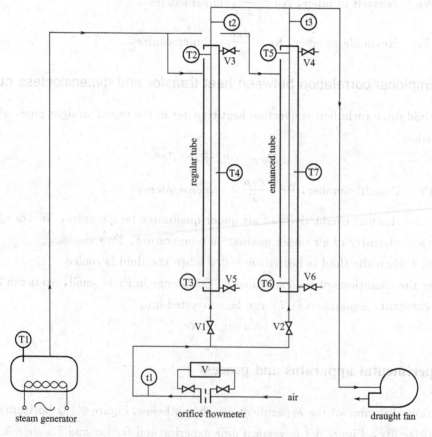

Figure 6-4　Measurement experiment of convective heat transfer coefficient of vertical tube

Heat exchange between air and steam: the steam from the steam generator enters the double-tube heat exchanger to exchange heat with the air drawn in by the draught fan. Non-condensable gas or non-condensed steam is discharged through the valves (V3 and V4). Condensate water is discharged by the drain valves (V5 and V6) into the water cup. The air is supplied by the draught fan. Its flow is automatically controlled by changing the fan speed using the frequency converter. The air passes through the orifice flowmeter and flows into the double-tube heat exchanger and is discharged from the draught fan outlet after heat exchange.

Note: the enhanced and regular tubes are chosen through the valves (V1 and V2) of the experimental apparatus, the knob on the instrument cabinet or the mouse on the computer interface.

The description of the symbols in the figure is shown in Table 6-1.

Table 6-1 Description of symbols

Symbol	Name	Unit	Remark
V	Air flow rate	m^3/h	Specification of copper tube $\phi 19mm \times 1.5mm$ Inner diameter of 16mm Effective length of 1020mm Range of air flow rate: 3 ~ 18m^3/h V1 and V2 are the valves in the pipeline. V3 and V4 are the valves for discharging non-condensable gas. V5 and V6 are the valves for discharging condensate water.
t1	Inlet air temperature	℃	
t2	Air temperature at the outlet of regular tube	℃	
t3	Air temperature at the outlet of enhanced tube	℃	
T1	Steam temperature in steam generator	℃	
T2	Temperature of outer copper wall at the air outlet of regular tube	℃	
T3	Temperature of outer copper wall at the air inlet of regular tube	℃	
T4	Steam temperature at the outside of regular tube	℃	
T5	Temperature of outer copper wall at the air outlet of enhanced tube	℃	
T6	Temperature of outer copper wall at the air inlet of enhanced tube	℃	
T7	Steam temperature at the outside of enhanced tube	℃	

Ⅳ Operation steps of virtual simulation experiment

1. Apparatus check for experiment

① Check whether the instrument, draught fan, steam generator and temperature measuring points are normal prior to experiment.

② Open the cover of the steam generator. Use a beaker to add water into the tank.

③ Pay attention to the liquid level indicator. Add water until the level exceeds 4/5 of the indicatior scale (as shown in Figure 6-6). Then cover the lid of the water tank.

④ Turn on the power switch (as shown in Figure 6-7), instrument switch and heat switch.

2. Experiment of regular tube

① Open the valve of the regular tube (as shown in Figure 6-8). Open two drain valves of non-condensable gas to appropriate position.

② Turn on the computer and open the software for experiment.

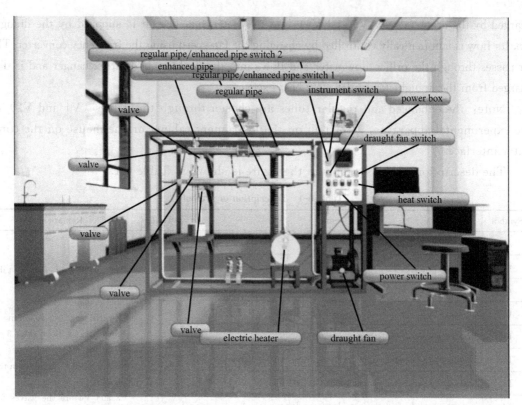

Figure 6-5 Virtual simulation diagram of measurement of convective heat transfer coefficient of horizontal tube

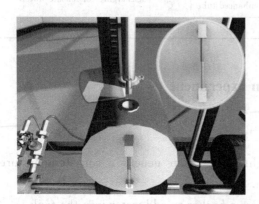

Figure 6-6 Supply water to steam generator

Figure 6-7 Turn on the instrument switch

Scan two-dimensional code: determination experiment of convective heat transfer coefficient

③ Click on the gas temperature of the heater. Click on the setting. Enter the temperature set value (such as 102) and click OK (as shown in Figure 6-9).

④ When a lot of non-condensable gas comes up and steam is discharged, it indicates that non-condensable gas is completely discharged.

⑤ Open the drain valve of condensate water of the regular tube until a little steam is discharged and condensate water is drained out

156　Chemical Engineering Principle Experiments and Virtual Simulation (Bilingual)

(as shown in Figure 6-10).

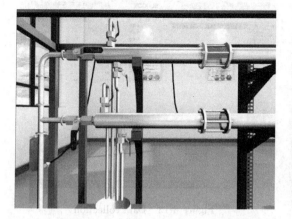

Figure 6-8　Open the valve of regular tube

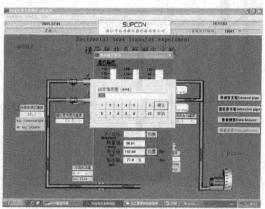

Figure 6-9　Control the temperature

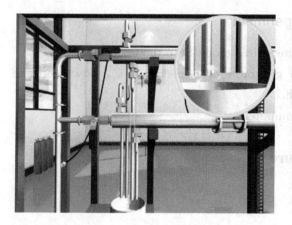

Figure 6-10　Steam production

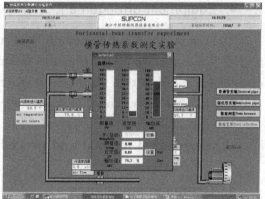

Figure 6-11　Set the flow rate

⑥ Turn on the fan switch on the control cabinet. Click the cold fluid flow rate on the software. Set the fluid flow rate (as shown in Figure 6-11).

⑦ The control cabinet and the software are switched to the regular tube.

⑧ When the flow is stable, record the data of the regular tube.

⑨ Rationally adjust the flow rate. Record several sets of data.

3. Experiment of enhanced tube

① Close the valve of the regular tube and the drain valve of condensate water.

② Open the valve of the enhanced tube (as shown in Figure 6-12).

③ Open the drain valve of condensate water of the enhanced tube to a suitable position. A little steam should be discharged meanwhile condensation water is drained out.

④ Replace the instrument cabinet and software for the experiment of enhanced tube.

⑤ Click the cold fluid flow rate on the software. Set the fluid flow rate.

⑥ After the flow and heat exchange is stable, collect the data (as shown in Figure 6-13).

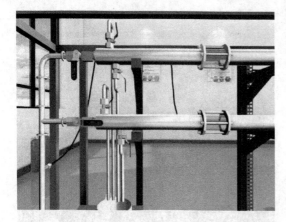
Figure 6-12 Open the valve of enhanced tube

Figure 6-13 Data collection

⑦ Rationally adjust the flow rate. Record several sets of data.

4. Equipment arrangement after experiment

① Turn off the steam generator. Reset the switch mode. Stop the instrument.

② When the temperature in the steam generator drops below 95℃, power off the draught fan and then turn off the main switch.

③ Close the valves and shut down the computer.

V Experimental method and procedure

1. Operating procedures

① Check whether the instrument, draught fan, steam generator and temperature measuring points are normal. Add water into the steam generator to 4/5 level of liquid level indicator.

② Turn on the main power and instrument power. Turn on the steam generator for heating. Meanwhile, slightly open the two valves for discharging non-condensable gas and control the temperature to be around 100℃ (as shown in Figure 6-6).

③ When large amount of steam raises through the valve (V3 or V4), then close the valve (V3 or V4), open the drain valve (V5 or V6) to constantly discharge the condensate water from the annulus space, in the meantime the discharge of non-condensed steam should be as minimized as possible.

④ Start the draught fan. Select the regular tube of the apparatus. The selected instrument and pipeline for measurement displayed on the computer must match the apparatus. Rationally distribute the flow rate by controlling the "Flow Setting" on the software, adjust the air volumetric flow rate to the maximum value for the experiment. Then gradually reduce the air flow rate. Generally choose 3~5 experimental flow points after the flow and heat exchange is stable, collect the data.

⑤ After the data of the regular tube are recorded, change the valves and select the enhanced tube. The selected instrument and pipeline for measurement displayed on the computer must match the apparatus. The method for measurement is the same as step ④.

⑥ At the end of the experiment, power off the steam generator. When the temperature in the steam generator drops below 95℃, shut off the draught fan and the main power. Clean the experiment equipment.

2. Attention

① Prior to the experiment, check whether the water level in the steam generator is normal. In case of low level, promptly add water.

② Read the data after the flow rate is adjusted and stable for at least 30 minutes.

③ Keep the stability of the rising steam and there should be steam continuously escaping from the vent in the experiment.

Ⅵ Experimental report

1. Fit the experimental equation $Nu = ARe^n$ and compare it with the recognized empirical equation of heat transfer (6-18). Compare the experimental results of regular and enhanced tubes for heat transfer.

2. Analyze the factors affecting heat transfer coefficient and the approach to enhance heat transfer.

3. Analyze the advantages and disadvantages of the enhanced tube heat transfer compared with the regular tube heat transfer.

Ⅶ Questions

1. What effect does the flow direction of cold fluid and steam have on the heat transfer in the experiment?

2. Is the density in the calculation of air mass flow rate consistent with that in the calculation of Reynolds number of air inside the heat exchange tube? What positions of density do they represent?

3. What impacts would occur if the condensate water is not discharged timely during the experiment? How to drain the condensate water timely? What impacts would the steam under different pressures in the experiment have on the correlation of α?

4. For this experiment, in order to improve the total heat transfer coefficient K, which effective method can be used? What is the most effective method?

5. The stability of which data is the main premise for recording experimental data, why?

Chapter 7 Filtration Experiment

I Experimental purposes and requirements

1. To understand and master the structure and operation of frame type filter process and vacuum filtration.
2. To learn the measurement methods of filtration constants K, q_e and virtual filter time τ_e and compressibility coefficient s.
3. To understand the impact of filtration pressure on filtration rate.

II Experiment principle

Filtration is an operation process in which suspension is separated into solid and liquid with the medium of porous materials, that is, under external forces, the liquid in the suspension flows through the layer of solid particles (i.e. filter cake layer) and the holes of porous medium but the solid particles are trapped to form a filter cake layer to achieve the separation of solid and liquid. Therefore, filtration actually shows the flow of fluid through a layer of solid particles. The thickness of this solid particle layer (filter cake layer) increases with the proceeding of filtration, so the filtration velocity decreases in the operation of filtration under constant pressure.

Production capacity of filtration equipment is expressed by filtration velocity, filtration velocity refers to the volume of filtrate flowing through the filter medium per unit filter area during per unit time. The basic factors influencing the filtering rate include: filter driving force (pressure difference) Δp, filter cake thickness L, the properties of suspension, suspension temperature and viscosity.

Filtration velocity

$$u = \frac{dV}{A\,d\tau} = \frac{dq}{d\tau} = \frac{A\Delta p^{(1-s)}}{\mu r C(V+V_e)} = \frac{A\Delta p^{(1-s)}}{\mu r' C'(V+V_e)} \tag{7-1}$$

where u—filtration velocity, m/s;

V—volume of filtrate flowing through filter medium, m³;

A—filter area, m²;

τ—filter time, s;

q—volume of filtrate flowing through filter medium per unit area, m³/m²;

Δp—filter pressure (gauge pressure), Pa;

s—compressibility coefficient of filter cake;

μ—filtrate viscosity, Pa·s;

r — specific resistance of filter cake, $1/m^2$;

C — volume of filter cake per unit volume of filtrate, m^3/m^3;

V_e — equivalent volume of filtrate of filter medium, m^3;

r' — specific resistance of filter cake, m/kg;

C' — mass of filter cake per unit volume of filtrate, kg/m^3.

When a certain suspension is filtered at constant pressure, Δp is constant at a work temperature, μ, r and C is constant too. Assuming that

$$K = \frac{2\Delta p^{(1-s)}}{\mu r C} \tag{7-2}$$

Then equation (7-1) can be converted into

$$\frac{dV}{d\tau} = \frac{KA^2}{2(V+V_e)} \tag{7-3}$$

where K — filtration constant, determined by material properties and filter pressure difference, m^2/s.

The equation below is obtained by the separation (7-3) of variable and integral:

$$\int_{V_e}^{V+V_e} (V+V_e) d(V+V_e) = \frac{1}{2} K A^2 \int_0^\tau d\tau \tag{7-4}$$

That is

$$V^2 + 2VV_e = KA^2 \tau \tag{7-5}$$

By changing the integral limit of equation (7-4) to $0 \sim V_e$ and $0 \sim \tau_e$, then the equation below can be obtained

$$V_e^2 = KA^2 \tau_e \tag{7-6}$$

By adding equation (7-5) and equation (7-6), the equation can be obtained as follow

$$(V+V_e)^2 = KA^2(\tau + \tau_e) \tag{7-7}$$

where τ_e — virtual filter time, equal to the time to filter out the filtrate with the volume of V_e, s.

Then by differentiating equation (7-7), the equation below can be obtained

$$2(V+V_e)dV = KA^2 d\tau \tag{7-8}$$

Then

$$\frac{\Delta \tau}{\Delta q} = \frac{2}{K}\bar{q} + \frac{2}{K}q_e \tag{7-9}$$

where Δq — filtrate volume per unit filter area measured every time (general equal distribution in experiment), m^3/m^2;

$\Delta \tau$ — corresponding time to measure the filtrate volume Δq every time, s;

$\bar{q}$ — average of two adjacent q values, m^3/m^2.

By fitting equation (7-9) into a straight line on the coordinate system taking $\Delta \tau/\Delta q$ as vertical axis and $\bar{q}$ as horizontal axis, the slope and intercept of the line can be obtained.

Slope

$$S = \frac{2}{K}$$

Intercept

$$I = \frac{2}{K} q_e$$

Then

$$K = \frac{2}{S}, \text{ m}^2/\text{s}$$

$$q_e = \frac{KI}{2} = \frac{I}{S}, \text{ m}^3$$

$$\tau_e = \frac{q_e^2}{K} = \frac{I^2}{KS^2}, \text{ s}$$

By changing the filtration pressure difference Δp, different values of K can be measured. By taking logarithm on both sides of equation (7-2), the following equation can be obtained

$$\lg K = (1-s)\lg(\Delta p) + B \qquad (7\text{-}10)$$

If B is a constant within the range of pressure difference in the experiment, the relationship of $\lg K \sim \lg(\Delta p)$ should present a line in the rectangular coordinate system. The slope is $(1-s)$ and the compressibility coefficient s of filter cake can be obtained.

III Experimental apparatus and process

1. Filtration of frame type filter press

This pressure filter is made of stainless steel tangential multi-layerpress, the experimental apparatus consists of a air compressor, a batching tank, a constant pressure tank and a frame type filter press. The experimental process is shown in Figure 7-1 and virtual simulation diagram is shown in Figure 7-2.

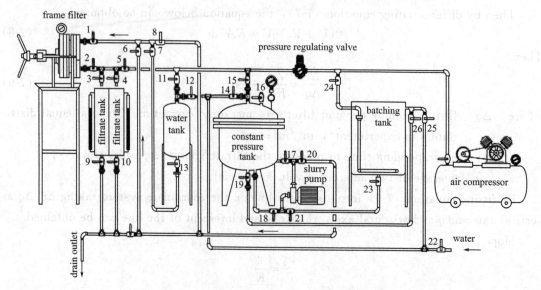

Figure 7-1 Filtration of frame type filter process

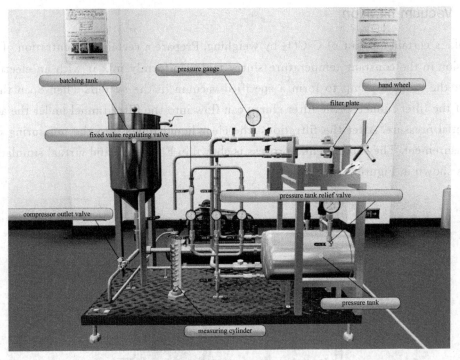

Figure 7-2 Diagram of virtual simulation filtration experiment of frame type filter process

After the $CaCO_3$ suspension with a certain concentration is prepared in the batching tank, feed it into the pressure tank using the slurry pump and stir it. Meanwhile, feed the filter slurry into the frame filter press using the compressed air (constant pressure) for filtration. The filtrate flows into the measuring cylinder for measurement and the compressed air is discharged.

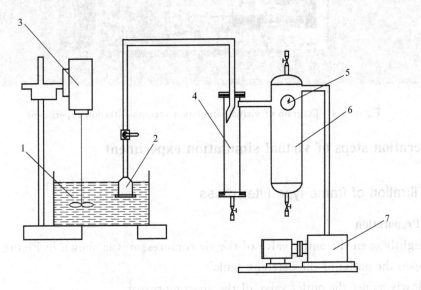

Figure 7-3 Vacuum filtration experiment flow diagram
1—Constant temperature slurry tank; 2—Filter funnel; 3—Stirring motor; 4—Measuring cylinder;
5—Vacuum pressure gauge; 6—Buffer tank; 7—Vacuum pump

2. Vacuum filtration

Take a certain amount of $CaCO_3$ by weighing. Prepare a certain concentration of $CaCO_3$ suspension in the constant temperature slurry tank. Uniformly mix it with an electric mixer. Start the vacuum pump to form a specified vacuum in the system. Then open the globe valve of the filter funnel so the filter slurry can flow into the filter funnel under the action of differential pressure. After the filtration, the clear liquid flows into the measuring cylinder for measurement. The experimental process is shown in Figure 7-3 and virtual simulation diagram is shown in Figure 7-4.

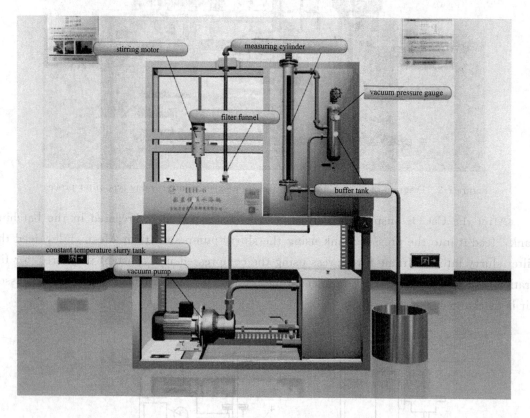

Figure 7-4 Diagram of virtual simulation vacuum filtration experiment

IV Operation steps of virtual simulation experiment

1. Filtration of frame type filter press

(1) Preparation

① Slightly open the outlet valve of the air compressor (as shown in Figure 7-5).

② Open the outlet of the batching tank.

③ Slowly adjust the outlet valve of the air compressor.

④ Install the frame filter (as shown in Figure 7-6). Properly install the filter plate, filter frame and filter cloth.

⑤ The filter cloth should cover the hole of the filter plate. The filter plate and filter frame should be separately installed. When it is pressed by screw compactor, prevent fingers from being crushed.

⑥ Open the relief valve of the pressure tank (as shown in Figure 7-7). Open the feed valve between the batching tank and pressure tank (as shown in Figure 7-8).

Scan two-dimensional code: filtration of frame type filter press

⑦ When the level of feed liquid in the pressure tank reaches 1/2 to 2/3 of observation mirror, close the feed valve.

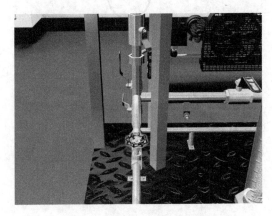

Figure 7-5 Slowly adjust the outlet valve of air compressor

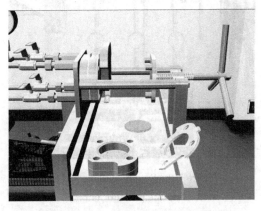

Figure 7-6 Installation of frame filter

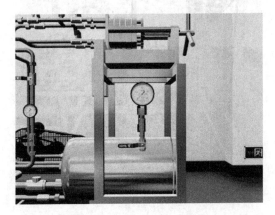

Figure 7-7 Open the relief valve of the pressure tank

Figure 7-8 Open the feed valve

(2) Filtration

① Open the fixed value regulating valve (0.1MPa) of the pressure tank to supply compressed air (as shown in Figure 7-9).

② Open the vent valve to continuously discharge air from the constant pressure tank. Spraying is not allowed.

③ Open one inlet valve before the feed frame (as shown in Figure 7-10).

④ Then slowly open the second valve while observe whether there is a spray from the frame type filter.

⑤ Open the globe valve of clear liquid after the discharge frame (as shown in Figure 7-11), filtrate flows out of the outlet of clear liquid.

⑥ Use a 500mL measuring cylinder to collect the filtrate. Record filter time when collecting 250mL filtrate every time (as shown in Figure 7-12).

⑦ Replace the measuring cylinder before the volume of the filtrate reaches 500mL.

Figure 7-9 Supply compressed air into the pressure tank

Figure 7-10 Open the inlet valve before the feed frame

Figure 7-11 Open the globe valve of clear liquid after the discharge frame

Figure 7-12 Record the filter time

(3) Change the filter pressure

① Close the frame filter inlet and outlet valves. Close the fixed value regulating valve.

② Open the relief valve of the pressure tank to release pressure.

③ Remove the filter frame, plate and cloth and clean them.

④ Prior to the next set of experiments, install the plate frame and turn down the relief valve.

(4) Filtration

① Open the second fixed value regulating valve (0.2MPa) of the pressure tank to supply compressed air.

② Open one inlet valve before the feed frame.

③ Then slowly open the second valve while observe whether there is a slurry spray from the frame type filter.

④ Open the globe valve of clear liquid behind the discharge frame. Filtrate flows out of the outlet of clear liquid. Collect the filtrate with the 500mL measuring cylinder.

⑤ Record filter time when collecting 250mL filtrate every time. Replace the measuring cylinder before the volume of the filtrate reaches 500mL.

⑥ Change the filter pressure to 0.25MPa. Repeat the operation.

(5) Completion

① Close the frame filter inlet and outlet valves.

② Close the fixed value regulating valve.

③ Open the relief valve of the pressure tank to release pressure.

④ Remove the filter frame, plate and cloth and clean them.

⑤ Close the outlet of the batching tank. Close the outlet valve of the air compressor.

2. Vacuum filtration

(1) Experiment preparation

① Check whether the electrical equipment works properly.

② Check whether the scale of measuring cylinder is marked correctly.

③ Turn on the power.

Scan two-dimensional code: vacuum filtration

(2) Operation

① Turn on the stirring motor for mixing (as shown in Figure 7-13). Adjust the motor speed between 0.08 and 0.10kr/min.

② Open the vent valve of the buffer tank (as shown in Figure 7-14).

Figure 7-13 Start the stirring motor   Figure 7-14 Open the vent valve of buffer tank

③ Start the vacuum pump (as shown in Figure 7-15) and change the pressure of buffer tank so the vacuum degree reaches 0.06MPa.

④ Note that a certain amount of liquid should be preserved in the measuring cylinder as the zero level.

⑤ Open the globe valve of the filter funnel (as shown in Figure 7-16).

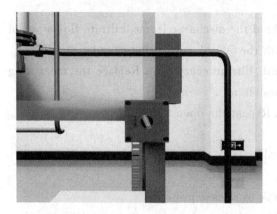 

Figure 7-15 Open the vacuum pump　　　　Figure 7-16 Open the globe valve of filter funnel

⑥ The slurry in the constant temperature slurry tank is driven by the pressure difference for filtration through the filter funnel. The clear liquid flows into the measuring cylinder.

⑦ Read the time with the stopwatch and record it every 2cm. Measure for 10 times (as shown in Figure 7-17).

Figure 7-17 Read and record the time　　　　Figure 7-18 Opening the drain valve

(3) Condition change

① Turn off the vacuum pump. Close the globe valve of the filter funnel.

② Open the drain valve to discharge the clear liquid from the measuring cylinder (as shown in Figure 7-18).

③ Keep a certain amount of liquid at the zero level with the same height as that in the first experiment.

④ Remove the filter funnel and cloth for cleaning.

⑤ Turn on the vacuum pump.

⑥ Control the buffer tank so the vacuum degree reaches 0.05MPa. Repeat the operation.

⑦ Then change the vacuum degree to 0.04MPa. Repeat the operation.

(4) Equipment arrangement after experiment

① Turn off the vacuum pump. Close the globe valve of the filter funnel.

② Turn off the stirring motor. Shut off the main power.

③ Open the drain valve of the measuring cylinder to discharge the clear liquid. Clean the experiment equipment.

V Experimental method and procedure

1. Filtration of frame type filter press

(1) Operating procedures

① Preparation: Open valve 25, close valve 25 after pouring about 25kg water into the batching tank. Then weigh 1.30kg $CaCO_3$ powder and add it into the batching tank.

② Start the air compressor (air compressor pressure limit can be set up to 0.2 MPa), use compressed air to mix liquid material by opening valve 24, open valve 23, 21, 17 and start the pump, pump all feed slurry into the constant pressure tank, then close valve 23, 21, now open valve 18, circulating mix slurry feeding in the constant pressure tank by the slurry pump.

③ Open valve 12 and valve 13, add water into the constant pressure water tank, close valve 12 and 13 after a specified period.

④ Adjust the constant pressure valve pressure index, generally control it to $0.05 \sim 0.10$MPa, then open valve 11, 15.

⑤ Open valve 1, 2 and 6, the filter pulp is pressed into the filter by compressed air, open valve 3 or 4 (only open one of them at measurement), put the filtrate into the tank, measure the filtrate volume and the corresponding time. When the filter press input pressure gauge reading is close to the constant pressure, the measurement is over. Open valve 9 or valve 10 to discharge filtrate.

⑥ Close valve 6, open valve 7, wash the filter cake with water.

⑦ Close valve 3 (or valve 4) and valve 7, open valve 5 and valve 8, blow residue by compressed air which helps the filter residue to fall off, then close valve 5 and valve 8. Discharge residue and wash filter cloth, if there is more material in the constant pressure tank, may open valve 26 or valve 19. The remaining material can be pressed into the batching tank with air or discharged, finally close valve 11 and valve 15, turn off the air compressor.

⑧ Clean the equipment at the end of the experiment.

(2) Attention

① Be sure to know the gas path of the whole process before starting the air compressor.

② Operating pressure should not be too large to prevent slurry from spraying out of the frame. Spread the filter cloth levelly.

③ Properly operate the vent valve and the pressure control valve.

2. Vacuum filtration

(1) Operating procedures

① Experiment preparation. Check whether the pipe joints leak and hoops are loose. Check whether the electrical equipment works properly. Check whether the scale of measuring cylinder is marked correctly. Confirm that the vacuum pump switches and constant temperature slurry tank heating switch are shut off. Connect the power (plug the power plug into the socket). Close all the valves on the apparatus.

② Preparation. Prepare the suspension containing 10%~30% (mass fraction) $CaCO_3$ in the constant temperature slurry tank. Use a balance to weigh the calcium carbonate in advance. Set the size of the constant temperature slurry tank as 490×325×115 (mm).

③ Mixing. Turn on the stirring motor for mixing. Properly control the mixing speed between 0.08~0.10kr/min.

④ Vacuum extraction. Open the vent valve of the buffer tank and start the vacuum pump. Then adjust the opening of the vent valve so that the vacuum degree of the system reaches the specified value (0.06MPa, 0.05MPa, 0.04MPa).

⑤ Filtration. Open the globe valve of the filter funnel so the filter slurry in the filter tank is driven by the pressure difference for filtration through the filter funnel. The clear liquid flows into the measuring cylinder. At this point, the vacuum pressure gauge indicates the vacuum degree after the filtration.

Stop the experiment if 10~12 readings are acquired for each vacuum degree.

⑥ Completion. Stop the vacuum pump; turn off the stirring motor and the main power. Open the drain valve of the measuring cylinder to completely discharge the clear liquid.

(2) Attention

① Vacuum degree is generally 0.04~0.06MPa. If vacuum degree is small, the opening of vent valve is large, vacuum pump would work with fresh air within generally three minutes; if vacuum degree is large, pipeline connection would be under much pressure resulting in gas access. A lot of white smoke would come out of the vacuum pump due to large vacuum degree if the pump has been used for a long time. At this time, the vacuum degree should be reduced or the pump should be stopped for cooling.

② The level of liquid in the measuring cylinder should not be higher than the inlet pipe of the buffer tank to prevent liquid from being pumped into the vacuum pump.

Ⅵ Experimental report

1. Calculate the filtration constants K, q_e and virtual filter time τ_e based on the data of vacuum filtration experiment.

2. Compare the values of K, q_e and τ_e under several differential pressures. Discuss the impact of differential pressure variation on the above parameters.

3. Fit the relationship curves of $\Delta\tau/\Delta q \sim \bar{q}$ and $\lg K \sim \lg \Delta p$ on the rectangular coordinate system. Calculate the value of s.

4. Analyze and discuss the experimental results.

Ⅶ Questions

1. What are the advantages and disadvantages of the frame type filter press? What occasions is it applied to?

2. What stages is frame type filter press operated at?

3. Why is the filter liquor a bit turbid at the beginning of filtration but clear after a period of time?

4. What are the principal factors affecting the filtration rate? How will the values of K, q_e and τ_e measured under a constant pressure change if the filter pressure is increased by double?

Chapter 8   Packed Column Absorption Experiment

I Experimental purposes and requirements

1. To understand the structure of packed absorption column and be familiar with its operation.

2. To observe the flooding of packed absorption column.

3. To determine the relationship curves of packing layer pressure drop Δp and superfical gas velocity u and measure the superfical gas velocity at flooding point.

4. To measure the volumetric absorption coefficient $K_Y a$ of the system of ammonia contained air-water.

II Experiment principle

Gas absorption is a typical mass transfer process.

1. Relationship between packing layer pressure drop Δp and superfical gas velocity u

When the gas passes through the dry packing layer (with a spraying density $L_0 = 0$), the relationship between pressure drop Δp and superfical gas velocity u is represented by the straight line in Figure 8-1 with a slope about 1.8, similar to the relationship between Δp and u when turbulent gas passes through the pipeline. Figure 8-1 shows that under the gas velocity below L and spraying with a certain density, the voidage decreases with the increasing liquid holdup resulting in the increase of pressure drop of the packing layer, moreover, the gas has no obvious impact on the flow of liquid film and the liquid holdup does not change with gas velocity, so the relationship between Δp and u is the same as that of dry packing material.

Under the spraying with a certain density, when the gas velocity moderately increases, liquid film thickens, "liquid blocking" would occur (above L point as shown in Figure 8-1) and the resistance of gas flowing through the packing layer greatly increases; if the gas velocity continues to increase, the dropping of spray liquid would be seriously blocked so the liquid would be extended from the packing surface to the whole packing layer, which is called "flooding" (above F point as shown in Figure 8-1) and the flow resistance of gas greatly increases. The F point in Figure 8-1 is the flooding point and the gas velocity u corresponding to it is called flooding gas velocity.

When superfical gas velocity u is above to flooding gas velocity, the pressure drop increases sharply and the operation is unstable, thus, the actual gas velocity is 60%~80% of flooding gas velocity.

(1) Superfical gas velocity u

Superfical gas velocity u refers to the gas velocity in the column represented by the ratio between its volumetric flow rate and column section area.

$$u = \frac{V'}{\Omega} \qquad (8-1)$$

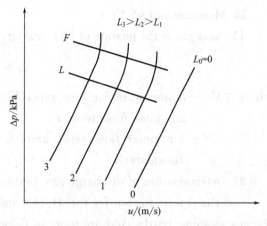

Figure 8-1 Relationship between packed column pressure drop Δp and superfical gas velocity u

where u—superfical gas velocity, m/s;
V'—gas volumetric flow rate in column, m³/s;
Ω—column section area, m².

The gas flow rate (air and ammonia) in the experiment needs to be measured with rotameter. Due to the inconsistency between the experimental measurement conditions and calibration conditions of rotameter, the readings of the rotameter must be corrected.

The corrected equation of gas rotameter is shown as follows

$$Q_S = Q_N \sqrt{\frac{p_N T_S}{p_S T_N}} \qquad (8-2)$$

where Q_S—actual gas flow rate, m³/h or L/h;
Q_N—reading of gas flowmeter, m³/h or L/h;
p_N, T_N—calibration status of calibration gas, $p_N = 1.013 \times 10^5$ Pa, $T_N = 293$ K;
p_S, T_S—absolute pressure and temperature of measured gas, Pa, K.

(2) Packing layer pressure drop Δp

The packing layer pressure drop Δp can be directly read out on the U-type differential pressure gauge 14.

2. Measurement of volumetric absorption coefficient $K_Y a$

(1) NH_3 absorption rate absorbed G_A is calculated according to the mass balance

$$G_A = V(Y_1 - Y_2) = L(X_1 - X_2) \qquad (8-3)$$

where G_A—absorption amount of NH_3 per unit time, kmol/h;
V—air flow rate, kmol/h;
L—water flow rate, kmol/h;
Y_1—mole ratio at gas inlet, kmol (A)/kmol (B);
Y_2—mole ratio at gas outlet, kmol (A)/kmol (B);
X_1—mole ratio at liquid outlet, kmol (A)/kmol (B);
X_2—mole ratio at liquid inlet, water in this experiment, $X_2 = 0$.

1) Measurement of Y_1

The inlet gas is the mixture of NH_3 and air, thus the mole ratio of NH_3 in the inlet gas is

$$Y_1 = \frac{V_A}{V} \tag{8-4}$$

where V_A —ammonia molar flow rate, kmol/h (obtained by correcting the readings of ammonia flowmeter);

V —air molar flow rate, kmol/h (obtained by correcting the readings of air flowmeter).

2) Determination of discharge gas (exhaust) composition Y_2

Use the pipette to transfer the H_2SO_4 standard solution with V_a mL of M_a (mol/L) into the gas washing bottle. Add appropriate deionized water and 3~5 drops of bromothymol blue. Connect it to the sampling exhaust pipe as shown in Figure 8-2. When the operation of the absorption column is stable, change the flow direction of the three-way cock so some tail gas at the top of the column can pass through the gas washing bottle and the ammonia in the tail gas is absorbed by H_2SO_4. The air flows out of the gas washing bottle into the wet type flowmeter for measurement. When the color of the solution in the bottle is changed from yellow (acidic) to green (neutral), it indicates the end of neutralization reaction (equivalent point). Quickly change the flow direction of the three-way cock. Prevent the gas from entering the gas washing bottle, or the solution will turn blue (alkaline).

$$Y_2 = \frac{n_{NH_3}}{n_{air}} \tag{8-5}$$

where n_{NH_3} —mole number of ammonia, mol;

n_{air} —mole number of air, mol.

Neutralization occurs between ammonia and sulfuric acid. Based on the amount of sulfuric acid, the amount of ammonia can be obtained as follows

$$n_{NH_3} = 2M_a V_a \times 10^{-3} \tag{8-6}$$

where V_a —standard H_2SO_4 solution volume, mL;

M_a —standard H_2SO_4 solution concentration, mol/L.

The air volume V_0 is obtained with the difference between readings of wet type gas flowmeter and its temperature T_0 is measured.

$$n_{air} = \frac{p_0 V_0}{R T_0} \tag{8-7}$$

where p_0 —air pressure through wet type flowmeter (indoor atmospheric pressure), Pa;

V_0 —air volume measured with wet type gas flowmeter, L;

T_0 —air temperature measured with wet type gas flowmeter, K;

R —gas constant, $R = 8314 N \cdot m/(mol \cdot K)$.

(2) Phase equilibrium constant m

The phase equilibrium relationship follows Henry's law system (generally refers to low concentration gas). The gas-liquid equilibrium relation

$$y^* = mx \tag{8-8}$$

The relationship between phase equilibrium constant m and system total pressure p and Henry coefficient E is as follows:

$$m = \frac{E}{p} \tag{8-9}$$

where E —Henry coefficient, Pa;

p —total system pressure (average pressure in the column), Pa.

According to the column top gauge pressure and the pressure difference between column top and bottom measured in the experiment, the average pressure in the column p can be obtained.

In this experiment, the relationship between Henry coefficient E and temperature T is as follows

$$\lg E = 11.468 - 1922/T \tag{8-10}$$

where T —liquid temperature (subject to the liquid temperature at the column bottom), K.

(3) Gas phase logarithmic mean concentration difference ΔY_m

$$\Delta Y_m = \frac{\Delta Y_1 - \Delta Y_2}{\ln \dfrac{\Delta Y_1}{\Delta Y_2}} \tag{8-11}$$

where
$$\Delta Y_1 = Y_1 - Y_1^* = Y_1 - mX_1$$
$$\Delta Y_2 = Y_2 - Y_2^* = Y_2 - mX_2$$

where Y_1^*, Y_2^* —gas phase equilibrium concentrations corresponding to liquid concentration X_1 and X_2, kmol (A)/kmol (B).

(4) Gas phase volumetric absorption coefficient $K_Y a$

Gas phase volumetric absorption coefficient $K_Y a$ is one of the main parameters reflecting the performance of a packed absorption column and its value is also an important basis for designing packed column. Low concentration gas absorption exists in this experiment, $Y \approx y$, $X \approx x$. Then

$$K_Y a = \frac{G_A}{\Omega h \Delta Y_m} \tag{8-12}$$

where $K_Y a$ —gas phase volumetric absorption coefficient, kmol/(m³ · h);

a —effective mass transfer area provided by packing layer per unit volume, m²/m³;

G_A —absorption amount of NH_3 per unit time, kmol/h;

Ω —column section area, m²;

h —packing layer height, m;

ΔY_m — gas phase mass transfer driving force, logarithmic mean concentration difference of gas phase.

Ⅲ Experimental apparatus and process

① The flow diagram and virtual simulation diagram of this experimental apparatus are shown in Figure 8-2 and Figure 8-3. The main equipment is the absorption column with an internal diameter of 70mm. Ceramic Rasching ring of 10mm×9mm×1mm is packed in the tower.

② The system is composed of water, ammonia and air. After the air containing inert gas is pumped out by the vortex air pump, its flow rate is measured with rotameter 4; ammonia is supplied with liquid ammonia cylinder and its flow rate is measured with rotameter 5; the flow rate of water as absorbent is measured with rotameter 3. Water is sprayed from the column top to packing layer to absorb the air containing ammonia rising from bottom to top. The solution at the column bottom flows out through the liquid seal tube. A liquid sample point is installed at the column bottom. The tail gas after absorption is discharged from the column top. Appropriate amount of tail gas is collected at the column top and its chemical composition is analyzed.

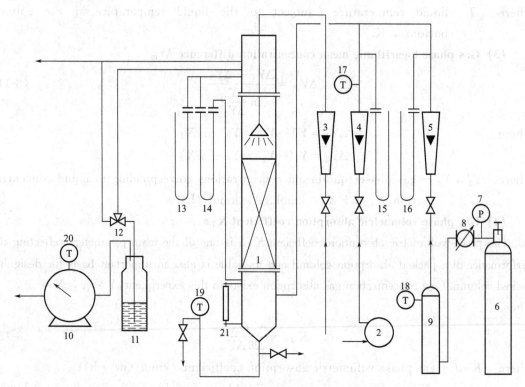

Figure 8-2 Packed column absorption flow diagram
1—Packed absorption column; 2—Vortex pump; 3—Rotameter (water); 4—Rotameter (air); 5—Rotameter (ammonia); 6—Liquid ammonia cylinder; 7—Ammonia pressure gauge; 8—Ammonia relief valve; 9—Ammonia buffer tank; 10—Wet type flowmeter; 11—Gas washing bottle; 12—Three-way cock; 13~16—U type differential pressure gauge; 17~20—Thermometer; 21—Liquid level gauge

Ⅳ Operation steps of virtual simulation experiment

1. Experiment when the density of water spraying is 0

① Read out the reading of Fortin barometer as the atmospheric pressure of this experiment.

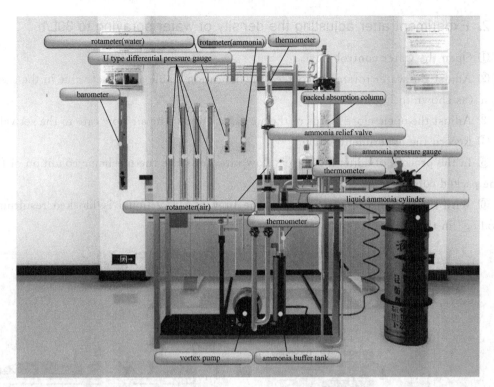

Figure 8-3 Virtual simulation diagram for packed column absorption

② Open the switch of the main power (as shown in Figure 8-4). Start the fan.

③ Open the air valve to supply air into the packed absorption column.

④ Adjust the air flow rate to the set value (as shown in Figure 8-5).

⑤ Record the air gauge pressure, column top gauge pressure, differential pressure between column top and bottom and air temperature.

Scan two-dimensional code: packed column absorption experiment

⑥ Adjust the air flow rate (distribute the flow rate) and record the experimental data after the air flow rate is adjusted.

Figure 8-4 Open the switch of main power

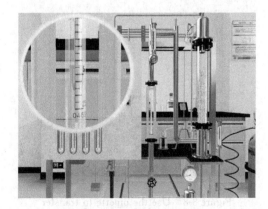

Figure 8-5 Set the air flow rate

2. Experiment after adjusting the density of water spraying to 30L/h

① Open the water control valve.

② Adjust the water rotameter to keep the indication stable at the set value in the case of 30L/h (as shown in Figure 8-6).

③ Adjust the open/close status of the air valve. Adjust the air flow rate to the set value.

④ Record the experimental data.

⑤ In the process of adjusting the air flow rate, observe the mechanic condition of fluid in the packed column.

⑥ With the increase in gas velocity, the flow of spray liquid is blocked resulting in "flooding" in the packed column (as shown in Figure 8-7).

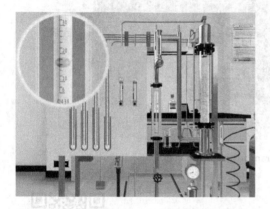

Figure 8-6 Set the water flow rate

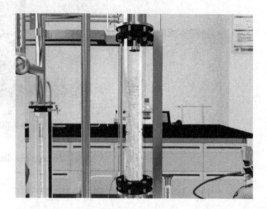

Figure 8-7 Flooding

3. Tail gas absorber assembly

① Tail gas absorber needs to be assembled before ammonia is supplied.

② Use a pipette to transfer the standard solution of sulfuric acid to the gas washing bottle (as shown in Figure 8-8).

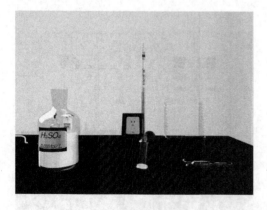

Figure 8-8 Use the pipette to transfer dilute sulfuric acid

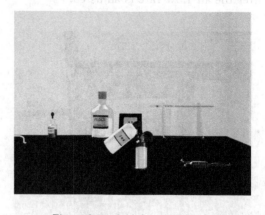

Figure 8-9 Add deionized water

③ Add two or three drops of bromothymol blue.

④ Add deionized water until it reaches the 2/3 level of the bottle (as shown in Figure 8-9).

⑤ Connect the inlet of the gas washing bottle to the sampling pipe of tail gas and connect the outlet to the wet type flowmeter.

⑥ Record the initial data of wet type flowmeter.

4. Measurement of volumetric absorption coefficient of the system containing ammonia, air and water

① Adjust the air valve to keep the indication of the air rotameter stable (such as $8m^3/h$).

② Adjust the water valve to keep the indication of the water rotameter stable (such as 30L/h).

③ Open the main valve of the liquid ammonia cylinder. Slowly adjust the relief valve of the cylinder.

④ The mole ratio of NH_3 in the inlet gas is around 3%. It is necessary to estimate the ammonia flow rate according to the air flow rate. Under the current condition, regulate the ammonia rotameter and keep the flow rate stable (as shown in Figure 8-10).

⑤ When the operation is stable, open the three-way cock and rotate it by 180° to supply tail gas into the gas washing bottle for composition analysis (as shown in Figure 8-11).

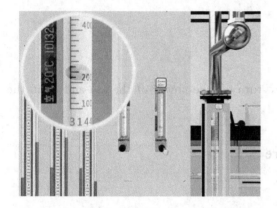

Figure 8-10 Set the ammonia flow rate

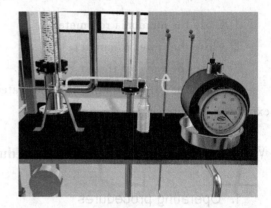

Figure 8-11 Tail gas analysis

⑥ When the color of the solution in the gas washing bottle is changed from yellow to green, it indicates that neutralization reaction reaches the equivalent point. At this point, the three-way cock should be immediately closed to prevent tail gas from being absorbed excessively.

⑦ If the supplied tail gas is excessively absorbed by accident, the solution in the bottle will become blue, then the composition of tail gas needs to be analyzed again.

5. Reading and recording of experimental data

① Record the data of the wet type flowmeter when the tail gas absorption is completed (as shown in Figure 8-12).

② Record the temperature of the air in the wet type flowmeter.

③ Record the ammonia temperature, ammonia gauge pressure, air gauge pressure, column top gauge pressure, differential pressure between column top and bottom, air temperature and column bottom liquid temperature (as shown in Figure 8-13).

④ Change the conditions for the experiment. Repeat the above operations.

Figure 8-12 Read the data of the wet type flowmeter   Figure 8-13 Read the gauge pressure

6. Equipment arrangement after experiment

① Close the ammonia rotameter.

② Close the water rotameter.

③ Close the valve of the air rotameter.

④ Close the valves of ammonia and water, turn off the power of the fan and clean the experiment equipment.

Ⅴ Experimental method and procedure

1. Operating procedures

(1) Measurement experiment of $\Delta p \sim u$

① When the density of water spraying is zero, start the fan, turn up the air flow rate (reasonable distribution of air flow rate), design the experiment record table and record the experimental data under different air flow rate.

② Open the absorber (water) control valve. Water flows into the packed column to wet the materials. Adjust the water flow rate to a set value. Turn up the air flow rate (rationally distributed in advance). Design the experiment record table and record the experimental data under different air flow rate meanwhile observe the phenomenon in the column (flooding) to obtain the flooding velocity (flooding point) under this water flow rate.

(2) Measurement experiment of volumetric absorption coefficient $K_Y a$

① Open the absorber (water) control valve and adjust the density of water spraying to the set value.

② Start the fan and regulate the flow rate to keep the indication of the rotameter stable at the set value.

③ Open the main valve of the liquid ammonia cylinder. Slowly adjust the relief valve of the cylinder so that its pressure can be stable at around 0.1~0.2MPa (operated by instructor).

④ Regulate the flow rate of ammonia rotameter (the ammonia concentration in the feed gas phase is about 3%) so that the indication is stable at the set value.

⑤ Prepare acidic absorption liquid to analyze the gas phase concentration of the column top, and accurately move a certain amount of absorbing liquid to the gas washing bottle.

⑥ After the operation is stable, open the three-way cock to supply the tail gas into the gas washing bottle for analyzing the composition of the tail gas. Record the readings of all flowmeters, thermometer and U-type differential pressure gauges and analyze the gas composition of the column top.

⑦ Change the operating condition (water flow rate). Repeat step ⑥.

⑧ Change the operating conditions again (air flow rate and ammonia flow rate). Repeat step ⑥.

⑨ At the end of the experiment, shut down the ammonia rotameter, then turn off the air and water rotameters and finally turn off the power of the fan and clean the experimental instruments.

2. Attention

① Many experimental data need to be recorded, design a complete data record sheet, cooperate with others to finish the experiment.

② In the process of experiment, especially when record the data, ensure the stability of air, ammonia and water flow rate.

VI Experimental report

1. Arrange the experimental data to obtain the data table of $\Delta p \sim u$. Plot $\Delta p \sim u$ based on the data table to obtain the flooding points. Compare the difference between the value and the experimental value.

2. Compare the change in volumetric absorption coefficient by changing gas flow rate with that by changing liquid flow rate to analyze the rationality of experimental results based on mass transfer theory and experimental process.

VII Questions

1. What is the practical significance of measuring the volumetric absorption coefficient $K_Y a$ and $\Delta p \sim u$?

2. How to determine the flow rate of water, ammonia and air in the experiment?

3. Why is ammonia transferred from the gas phase to the liquid phase in the experiment?

4. What are the impacts of the changes in gas and liquid flow rate on the $K_Y a$ in the experiment in theory?

5. Which operating conditions affect the stability of the experiment system?

6. What are the main experimental errors? What methods can be utilized to reduce the errors?

Chapter 9 Sieve-plate Column Distillation Experiment

I Experimental purposes and requirements

1. To review the basic theories of plate column for distillation and understand the basic structure, typical process and ancillary equipment of plate column.
2. To understand the design of plate column and grasp the operation methods.
3. To know the concept of sensitive plate and learn the method of judging whether the distillation system is stable.
4. To calculate the overall efficiency of the sieve-plate distillation column at total reflux and partial reflux.
5. To design a process for ethanol separation from ethanol solution in continuous distillation, and separate an certain amount of raw material within a period of time.

II Experiment principle

The plates of the sieve-plate distillation column are the place where gas and liquid come into contact. Rising steam produced at column bottom gradually contacts through heat and mass transfer with the falling liquid from column top. The falling liquid is partially evaporated several times and its heavy components are gradually increased. The rising steam is partially condensed several times and its light components are gradually increased. Thus, the mixture is separated to a certain degree.

1. Overall efficiency E_T at total reflux

$$E_T = \frac{N_T - 1}{N_P} \times 100\% \tag{9-1}$$

where N_T—the number of theoretical plates required to complete a certain separation, including distillation still;

N_P—the number of practical plates required to complete a certain separation, $N_P = 7$ in this experiment.

In the operation at total reflux, the operation line is a diagonal line on the $x \sim y$ graph in Figure 9-1. By stepping the material composition at column top and bottom between the operating line and equilibrium line, the number of theoretical plates N_T can be obtained.

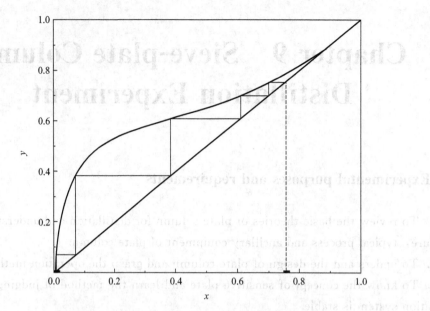

Figure 9-1　Graph for the determination of the number
of theoretical plates at total reflux

2. Overall efficiency E_T' at partial reflux

(1) Operating line equation of the rectifying section

$$y_{n+1} = \frac{R}{R+1}x_n + \frac{x_D}{R+1} \qquad (9\text{-}2)$$

where　y_{n+1} —mole fraction of the rising steam of rectifying section $n+1$ plate;

　　　　x_n —mole fraction of the falling liquid of rectifying section n plate;

　　　　R —reflux ratio, $R = L/D$;

　　　　x_D —mole fraction of the liquid product at column top.

In the experiment, the quantity of reflux is measured with reflux rotameter. But cold liquid is generally used for reflux in the operation, so the actual quantity of reflux needs to be corrected.

$$L = L_0 \left[1 + \frac{c_{pD}(t_D - t_R)}{r_D} \right] \qquad (9\text{-}3)$$

where　L_0 —reading on the reflux rotameter, mL/min;

　　　　L —actual quantity of reflux, mL/min;

　　　　t_D —liquid temperature at column top, ℃;

　　　　t_R —reflux temperature, ℃;

　　　　c_{pD} —specific heat of the column top reflux at average temperature $(t_D + t_R)/2$, kJ/(kg·K);

　　　　r_D —latent heat of vaporization of the column top reflux composition, kJ/kg.

The flow rate of product D can be measured with rotameter. Due to the same composition and temperature of product flow rate D and quantity of reflux L, the reflux ratio R can be directly obtained from the ratio between them.

$$R = \frac{L}{D} \tag{9-4}$$

where D—reading on the rotameter of product, mL/min.

According to the sampling analysis of column top in the experiment, x_D can be obtained, and the readings of reflux and product flow rate from rotameter L_0, D as well as the reflux temperature t_R and column top liquid temperature t_D can be measured. Check the attached table to know c_{pD} and r_D. Based on equation (9-3) and equation (9-4), the reflux ratio R can be obtained. By substituting it into equation (9-2), the operating line equation of rectifying section can be obtained.

(2) feed line (q-line) equation

$$y = \frac{q}{q-1}x - \frac{x_F}{q-1} \tag{9-5}$$

where q—liquid fraction of feed;

x_F—liquid feed composition, mole fraction.

$$q = \frac{\text{heat required for 1kmol feed liquid changing into saturated steam}}{\text{latent heat of 1kmol feed liquid}} = 1 + \frac{c_{pF}(t_S - t_F)}{r_F} \tag{9-6}$$

where t_S—bubble point temperature of the liquid feed, ℃;

t_F—liquid temperature, ℃;

c_{pF}—specific heat of the liquid feed at average temperature $(t_S + t_F)/2$, kJ/(kg·K);

r_F—latent heat of vaporization of the liquid feed composition, kJ/kg.

Make sampling analysis to obtain the distillate composition x_D, bottom composition x_W and feed composition x_F and check the attached table to know t_S, c_{pF} and r_F. By substituting them into equation (9-6), the q-line equation can be obtained.

(3) Calculation of the number of theoretical plates

According to the operating line equation of rectifying section and q-line equation, and measured column top composition x_D, column bottom composition x_W and feed composition x_F, q-line and operating line of stripping section can be drawn on the $x \sim y$ graph. Then the number of theoretical plates N_T can be obtained with $x \sim y$ graphic method (as shown in Figure 9-2).

(4) Overall efficiency E'_T at partial reflux

According to the number of theoretical plates N_T obtained above, the overall efficiency E'_T at partial reflux can be obtained from equation (9-1).

$$E'_T = \frac{N_T - 1}{N_P} \times 100\% \tag{9-7}$$

where N_T—the number of theoretical plates required to complete a certain separation, including distillation still;

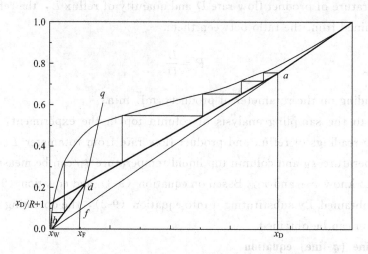

Figure 9-2 Graph for the determination of the number
of theoretical plates at partial reflux

N_P —the number of practical plates required to complete a certain separation, N_P =7 in this experiment.

III Experimental apparatus and process

The distillation column is composed of sieve plate column still, column body (7 plates), total condenser, feeding system, reflux system, storage tank (raw material, product and residue), product discharge pipe, residue discharge pipe, cooling water rotameter, centrifugal pump as well as measurement and control instruments. The flow diagram of this experimental apparatus is shown in Figure 9-3 and the virtual simulation diagram is shown in Figure 9-4.

The sieve plate column has an internal diameter of 68mm and 7 plates among which 5 plates are at rectifying section and 2 plates at stripping section; the space between the plates at rectifying section is 150mm and that at stripping section is 180mm; the sieve aperture is 1.5mm and the sieve holes are arranged in equilateral triangle shape with a space of 4.5mm and hole number of 104. This apparatus is electrically heated, equipped with three spiral tube heaters with a rated power of 3kW. Sampling points of liquid, product and residue are installed on the apparatus (A, B, C in the figure).

IV Operation steps of virtual simulation experiment

1. Experiment preparation

① The feed liquid prepared according to the required concentration is fed to the storage tank to the top of the glass level gauge.

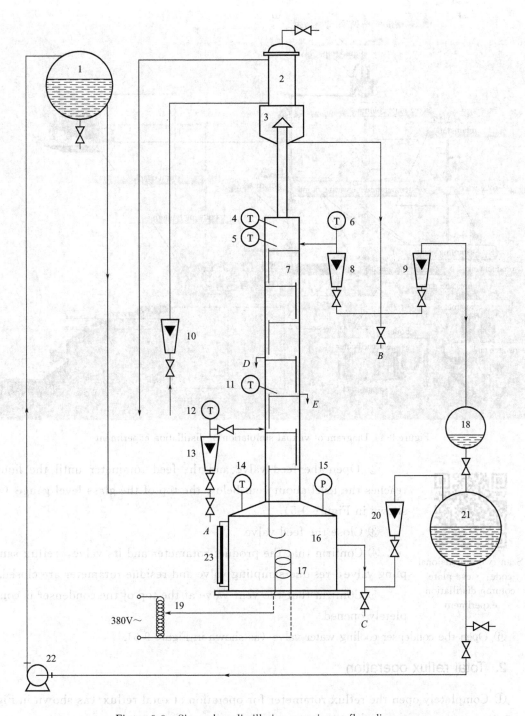

Figure 9-3　Sieve-plate distillation experiment flow diagram

1—Storage tank；2—Condenser；3—Liquid trap；4～6,11,12,14—Temperature measuring point；
7—Sieve plate column；8～10,13,20—rotameter；15—Pressure gauge；16—Column still；
17—Reboiler；18—Product tank；19—Voltage regulator；21—Residue tank；
22—Infusion pump；23—Level gauge；$A \sim E$—Sampling point

Chapter 9　Sieve-plate Column Distillation Experiment

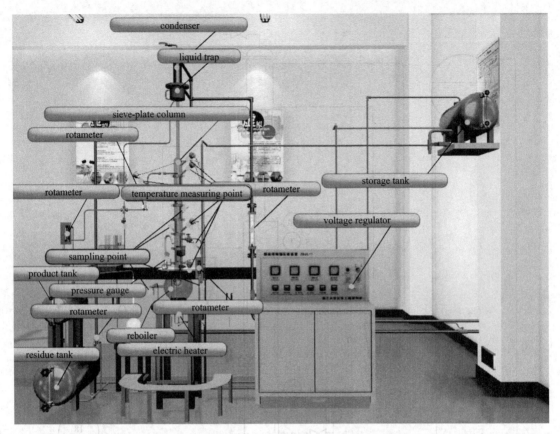

Figure 9-4　Diagram of virtual simulation of distillation experiment

Scan two-dimensional code: sieve-plate column distillation experiment

② Open the feed valve and the feed rotameter until the liquid reaches the level about 1cm below the top of the glass level gauge (as shown in Figure 9-5).

③ Close the feed valve.

④ Confirm that the product rotameter and its valve, reflux sampling valve, residue sampling valve and residue rotameter are closed.

⑤ Confirm that the vent valve at the top of the condenser is completely opened.

⑥ Open the condenser cooling water valve (as shown in Figure 9-6).

2. Total reflux operation

① Completely open the reflux rotameter for operation at total reflux (as shown in Figure 9-7).

② Turn on the main power of the instrument cabinet.

③ Adjust the voltage to 190~210V and keep it constant until the residual liquid boils (as shown in Figure 9-8).

④ Observe the temperature of the column plates every five minutes. When the temperature of the sensitive plate is nearly constant, the operation is stable.

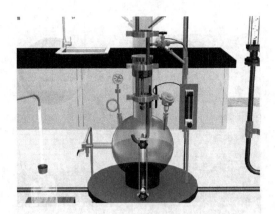

Figure 9-5   Open the feed rotameter

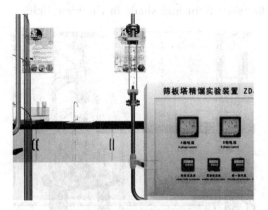

Figure 9-6   Open the condenser cooling water valve

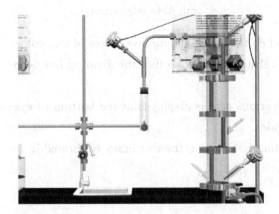

Figure 9-7   Completely open the reflux rotameter

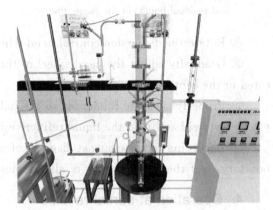

Figure 9-8   Residual liquid boiling

3. Data recording and sample analysis

① Record the readings of the rotameter for cooling water and rotameter for reflux as well as the data on the instrument cabinet.

② Take out the samples of distillate and residual liquids at the same time and analyze the compositions (as shown in Figure 9-9).

③ Place disposable tubes at the sampling points of feed liquid to take samples and open the sampling valve. Add the feed liquid until the level reaches 2/3 tube.

④ Cool the sample to the room temperature and analyze the concentration with Abbe refractometer (as shown in Figure 9-10).

⑤ After the solution is cooled, use a disposable pipette to take the liquid and drop it on the polished surface of the refraction prism seat. The liquid layer should be uniform in full view without bubbles.

⑥ Cover the prism and tighten it with the hand wheel, open the light screen and close the mirror.

⑦ Rotate the hand wheel of the refractive index and find the location of the dividing line between light and shade in the view field.

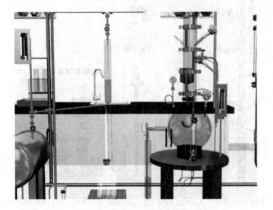

Figure 9-9　Take out the samples of distillate and residual liquids at the same time

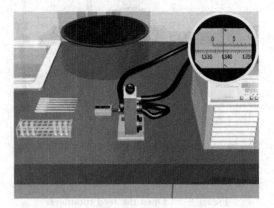

Figure 9-10　Composition analysis with Abbe refractometer

⑧ Rotate the dispersion control hand wheel so that the dividing line is free of any color.

⑨ Gradually adjust the hand wheel of the refractive index so that the dividing line is located in the center of the reticle.

⑩ Properly turn the condenser so the indications can be displayed at the bottom of eyepiece view field and read the liquid refractive index.

⑪ According to the contrast diagram of concentration～refractive index calibrated in laboratory, get the concentration of all samples.

4. Operation at partial reflux

① Open the feed rotameter and adjust the flow rate to 200～250mL/min.

② Open the valve and adjust the product rotameter to set the reflux ratio to 3～5 (as shown in Figure 9-11).

③ Adjust the residual rotameter (as shown in Figure 9-12) to keep the level of residual at the column bottom stable.

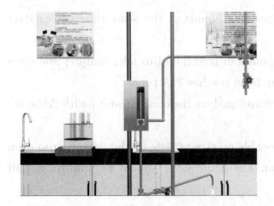

Figure 9-11　Adjust the product rotameter

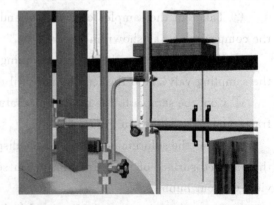

Figure 9-12　Adjust the residual liquid rotameter

④ When the level of residual liquid and temperature of the sensitive plate are stable, it indicates that the operation at partial reflux is stable.

⑤ Record the data on the instrument cabinet.

⑥ Record the readings of feed rotameter, reflux rotameter, product rotameter and residue rotameter.

⑦ Take the samples of feed liquid, distillate liquid and residual liquid for analysis with Abbe refractometer.

⑧ According to the contrast diagram of concentration ~ refractive index calibrated in laboratory, get the concentration of all samples.

5. Equipment arrangement after experiment

① Close the feed liquid rotameter and feed valve.
② Close the rotameter and valve involving the product.
③ Close the residual liquid rotameter.
④ Turn the knob of the pressure regulator to zero position and turn off the main power.
⑤ Close the condensate water valve after ethanol steam is completely condensed.

V Experimental method and procedure

1. Operating procedures

① Prepare the solution according to the required concentration (generally $x_F = 0.1$) and feed about 9L liquid into the column until the level reaches the top of the glass gauge. If the preparation is completed before conducting the experiment, determine the composition of the feed.

② Close the feed and discharge valves and the sampling valve. Open the vent valve at the top of the condenser to its full open position. Slightly open the condensate water valve. Fully open the reflux rotameter for operation at total reflux.

③ Turn on the main power of the instrument cabinet. Rotate the knob of the voltage regulator to adjust the voltage to the desired value for heating and keep it constant.

④ When the residual liquid begins to boil, turn up the condensate water valve until the reading of the rotameter reaches the maximum value and keep it constant.

⑤ After the voltage for heating and condensate water volume are stable, observe the temperature of the column plates every five minutes. When the temperature of the sensitive plate is nearly constant, the operation is stable. Take the samples of distillate and residue to analyze their composition. Measure the flow rate of reflux. Then measure the temperatures of the distillation column.

⑥ For the partial reflux operation, open the feed valve to adjust the feed rate to 200~250mL/min and adjust the flowmeter to set the reflux ratio to 3~5. By regulating the rotameter of residual liquid, the level of the residual liquid at the column bottom should remain unchanged. When the level of the residual liquid and temperature of the sensitive plate are stable, it indicates that the operation at the partial reflux is stable. Take the samples of feed, distillate and residue to analyze their composition. Record the relevant data. Measure the flow rate of feed, distillate, residue and reflux. Then measure the six temperatures of the distillation column.

⑦ Upon completion, close the flow control valves of feed liquid, distillate and residual liquid and then turn the knob of the voltage regulator to the zero position. Turn off the main power, when the ethanol steam is completely condensed, close the condensate water valve and clean the apparatus.

2. Attention

① At the beginning of the experiment, supply cooling water into the condenser at the top of the column then heat the column bottom, and reverse the steps at the end of the experiment.

② The position of the liquid in the column bottom should always be higher than the electrode of the heating vessel.

Ⅵ Experimental report

1. Keep the raw data of column top and bottom temperature, composition and readings of instruments in the list.
2. Calculate the number of theoretical plates with graphic method under the conditions of total reflux and partial reflux.
3. Calculate the overall efficiency.
4. Analyze and discuss the phenomena observed during the experiment.

Ⅶ Questions

1. What are the factors affecting the stability of distillation? What should be given attention to maintain the stability of the column? How to determine the operation stability of column?
2. What impact does the changed heating power have on the separation in the condition of total reflux?
3. What impact does the cold reflux at column top have on the amount of reflux liquid in the column? How to improve it?
4. How to make corrections in the calculation when the ethanol-water solution flow rate is measured with rotameter?
5. What are the main factors influencing the efficiency of the whole column?

Chapter 10 Extraction Experiment

I Experimental purposes and requirements

1. To understand the basic structure, operation methods and extraction process of rotating disc column and pulsed column.

2. To observe the flowing of light and heavy phases in the extraction column when there are changes in rotary speed of rotating disc column or pulse intensity of pulsed column (pulse amplitude and frequency), and understand the main factors affecting the extraction and analyze the impact of extraction conditions on extraction process.

3. To calculate the number of mass transfer units, height of a transfer unit and volumetric mass transfer coefficient K_{YV} and determine the relationship between height of a transfer unit and operating variables of pulsed extraction.

4. To calculate the extraction efficiency η.

II Experiment principle

Extraction is one of the important operations to separate and purify a substance and it is a unit operation to separate components by using the difference of solubility between components of a mixture in the additive solvent. In the operation of liquid-liquid extraction, two liquids flow reversely in the column. One liquid acts as a disperse phase and passes through the other liquid as a continuous phase in the form of droplets. The concentrations of the two liquid phases are continuously differentiated in the column and the separation between the two liquid phases is achieved at both ends of the column depending on the density difference. When the light phase acts as the dispersed phase, the phase interface appears in the upper part of the column; on the contrary, the phase interface appears in the lower part of the column. In this experiment, the light phase is the dispersed phase, and the phase interface appears in the upper part of the column.

When calculating the height of differential countercurrent extraction column, the mass transfer unit method is adopted. That is, characteristics are represented by the number of mass transfer units and height of a transfer unit. The number of mass transfer units indicates the degree of difficulty in the process of separation and the height of a transfer unit indicates the mass transfer performance of equipment.

1. Basic symbols of extraction operation (as shown in Table 10-1)

Table 10-1 basic symbols of extraction operation

Name	Symbol	Flow unit	Composition symbol
Feed liquid	F	kg/s	X_F or x_F
Raffinate	R	kg/s	X_R or x_R
Extraction agent	S	kg/s	Y_S or y_S
Extraction phase	E	kg/s	Y_E or y_E

2. Mass balance of extraction operation

As shown in Figure 10-1, Figure 10-2, the composition of each component in the extraction calculation is associated with the operating line equation. P (X_R, Y_S) and Q (X_F, Y_E) of the operating line equation correspond to the upper and lower parts of the column.

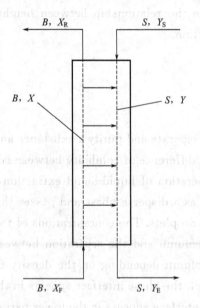 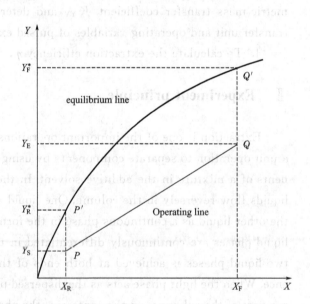

Figure 10-1 Mass balance of extraction operation　　Figure 10-2 Average driving force calculation

When the first solvent B is completely immiscible with the extraction agent S, the operating line of the extraction process is linear on the $X \sim Y$ coordinates, and the equation is as follows.

$$\frac{Y_E - Y_S}{X_F - X_R} = \frac{Y - Y_S}{X - X_R} \tag{10-1}$$

The equation below is obtained from the above equation

$$Y - Y_S = m(X - X_R)$$

where

$$m = \frac{Y_E - Y_S}{X_F - X_R}$$

The amount M of pure material A extracted from the first solvent per unit time can be determined by mass balance

$$M = B(X_F - X_R) = S(Y_E - Y_S) \qquad (10\text{-}2)$$

3. Mass transfer in the extraction process

Unbalanced extraction phase and raffinate phase contact in any section of the column and mass transfer is held between the two phases. Substance A transfers from raffinate phase into extraction phase through diffusion. The limit of this process is to achieve phase equilibrium, and the phase equilibrium relationship is

$$Y^* = kX \qquad (10\text{-}3)$$

k is the partition coefficient. Only in the simple system, k is a constant. Under normal circumstances, k is a variable. In this experiment, the system (kerosene-benzoic acid-water) equilibrium data are shown in Table 10-2 and can be used to obtain the relationship between X and Y.

Table 10-2 Equilibrium data of kerosene-benzoic acid-water system at room temperature

y (mass fraction)	x (mass fraction)	y (mass fraction)	x (mass fraction)
0.00972	0.0128	0.1144	0.1786
0.0195	0.0199	0.1301	0.2348
0.0354	0.0270	0.1782	0.4230
0.0683	0.0817	0.2195	0.6550
0.0725	0.0990	0.2220	0.6330
0.1010	0.1494		

where, x—mass fraction of benzoic acid in oil phase; y—mass fraction of benzoic acid in water phase.

Low concentration exists in this experiment, so $Y \approx y$, $X \approx x$.

The degree of deviation from the equilibrium composition is the driving force in the process of mass transfer. At the top of the column, the driving force in Y phase is the line segment PP'

$$\Delta Y_R = Y_R^* - Y_S \qquad (10\text{-}4)$$

The driving force at the lower part of the column in Y phase is the line segment QQ'

$$\Delta Y_F = Y_F^* - Y_E \qquad (10\text{-}5)$$

In the condition that the operating line and equilibrium line are linear, the average driving force of mass transfer is

$$\Delta Y_m = \frac{\Delta Y_F - \Delta Y_R}{\ln \dfrac{\Delta Y_F}{\Delta Y_R}} \qquad (10\text{-}6)$$

The mass transfer kinetic equation of material A flowing from raffinate phase into extraction phase is

$$M = K_Y A \Delta Y_m \qquad (10\text{-}7)$$

where K_Y —mass transfer coefficient, $kg/(m^2 \cdot s)$;
A —phase contact surface area, m^2.

The phase contact surface area A in the extraction column in this equation cannot be determined, so an alternative method is usually used.

The phase contact surface area A can be represented by the equation below.

$$A = aV = a\Omega h \tag{10-8}$$

where a —mass transfer area per unit effective volume of column, m^2/m^3;
V —volume of effective operating section in the extraction column, m^3;
Ω —section area of extraction column, m^2;
h —height of the operation section in the extraction column, m.

At this point

$$M = K_Y a V \Delta Y_m = K_{YV} V \Delta Y_m \tag{10-9}$$

where $K_Y a = K_{YV}$ —volumetric mass transfer coefficient, $kg/(m^3 \cdot s)$.

According to equation (10-2), equation (10-7), equation (10-8) and equation (10-9), the following equation can be obtained.

$$h = \frac{S}{K_{YV}\Omega} \times \frac{Y_E - Y_S}{\Delta Y_m} = H_{OE} N_{OE} \tag{10-10}$$

where $H_{OE} = \dfrac{S}{K_{YV}\Omega}$ —height of a mass transfer unit in extraction phase;

$N_{OE} = \dfrac{Y_E - Y_S}{\Delta Y_m}$ —number of mass transfer units in extraction phase.

K_Y, K_{YV} and H_{OE} represent the characteristics of mass transfer. The greater the K_Y and K_{YV} are, the smaller the H_{OE} is, and the faster the extraction is.

$$K_{YV} = \frac{M}{V \Delta Y_m} = \frac{S(Y_E - Y_S)}{V \Delta Y_m} \tag{10-11}$$

4. Extraction efficiency

$$\eta = \frac{\text{amount of component A extacted by extractant}}{\text{amount of component A in feed solution}} \times 100\%$$

So

$$\eta = \frac{S(Y_E - Y_S)}{B X_F} \times 100\% \tag{10-12}$$

Or

$$\eta = \frac{B(X_F - X_R)}{B X_F} \times 100\% = \left(1 - \frac{X_R}{X_F}\right) \times 100\% \tag{10-13}$$

5. Mass flow rate and composition

① Mass flow rate of the first solvent B

$$B = F(1-x_F) = V_F \rho_F (1-x_F) \tag{10-14}$$

where B —mass flow rate of the solvent, kg/h;

F —mass flow rate of feed liquid, kg/h;

V_F —volumetric flow rate of feed liquid, m³/h;

ρ_F —density of feed liquid, kg/m³;

x_F —the content A in feed liquid, kg/kg.

Liquid flowmeter calibration: V_F is calculated from equation (10-15)

$$V_F = V_N \sqrt{\frac{\rho_0(\rho_f - \rho_F)}{\rho_F(\rho_f - \rho_0)}} \approx V_N \sqrt{\frac{\rho_0}{\rho_F}} \tag{10-15}$$

where V_N —reading of rotameter, mL/min or m³/h;

ρ_f —rotor density, kg/m³;

ρ_0 —density of water at 20 ℃, kg/m³.

So

$$B = V_N \sqrt{\rho_0 \rho_F}(1-x_F) \tag{10-16}$$

② Mass flow rate of extraction agent (water) S

$$S = V_S \rho_S \tag{10-17}$$

where V_S —volumetric flow rate of extraction agent, m³/h;

ρ_S —density of extraction agent, kg/m³。

③ Composition of feed solution and raffinate x_F and x_R

For the system composed of kerosene, benzoic acid and water, the composition of feed liquid x_F, raffinate phase x_R and extraction phase y_E can be determined using the acid-base neutralization titration method, i.e. the mass fraction of benzoic acid. y_E can also be calculated with the above mass balance. The specific steps are as follows.

Use the pipette to take V_1 mL sample and add 1~2 drops of indicator. Use the NaOH solution with a concentration of N_b to the end point of titration. If V_2 mL NaOH solution is used, the molar concentration of benzoic acid N_a in the sample is

$$N_a = \frac{V_2 N_b}{V_1} \tag{10-18}$$

Then

$$x_F = \frac{N_a M_A}{\rho_F} \tag{10-19}$$

where M_A —molecular weight of solute A, g/mol, the molecular weight of benzoic acid is 122 g/mol;

ρ_F —solution density, g/L.

x_R is determined in the same way.

$$x_R = \frac{N'_a M_a}{\rho_R} \tag{10-20}$$

$$N'_a = \frac{V'_2 N_b}{V'_1} \tag{10-21}$$

where V_1', V_2'—the sample volume and volume of NaOH solution for titration.

III Experimental apparatus and process

1. Rotating disc column

Rotating disc column is the main equipment for the experiment. The column body is made of glass tube with an internal diameter of 50mm. The motor at the top of the column is connected to a shaft. The disc is fixed on the shaft. The circular ring is fixed on the column wall. The circular ring and disc are arranged in a staggered manner. The flow diagram of rotating disc extraction is shown in Figure 10-3, and the virtual simulation diagram is shown in Figure 10-4.

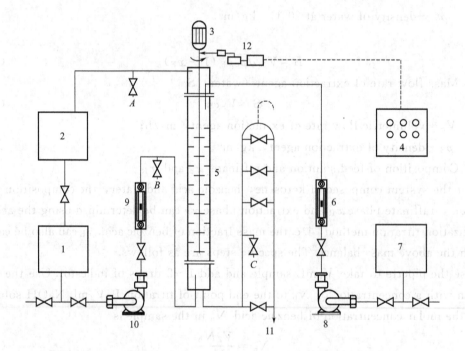

Figure 10-3 Rotating disc extraction experiment flow diagram

1—Storage tank (benzoic acid-kerosene); 2—Collection tank (raffinate); 3—Motor; 4—Control cabinet;
5—Rotating disc column; 6,9—Rotameter; 7—Extractant storage tank (water);
8,10—Pump; 11—Discharge (extraction liquid) tube;
12—Speed measuring meter; A~C—Sampling point

2. Pulsed extraction column

Pulsed extraction column is the main equipment for the experiment. The column body is made of glass tube with an internal diameter of 50mm, stuffed with stainless steel mesh. The flow diagram of pulsed extraction is illustrated in Figure 10-5 and the virtual simulation

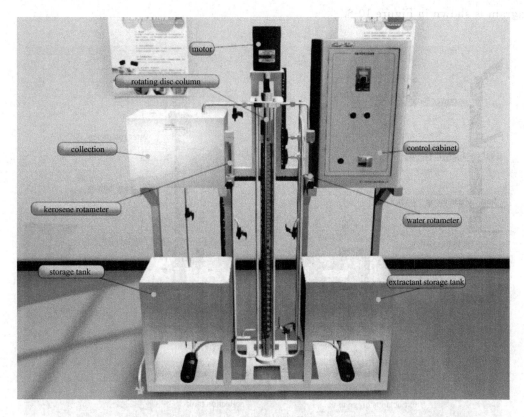

Figure 10-4　Diagram of virtual simulation of rotating disc extraction experiment

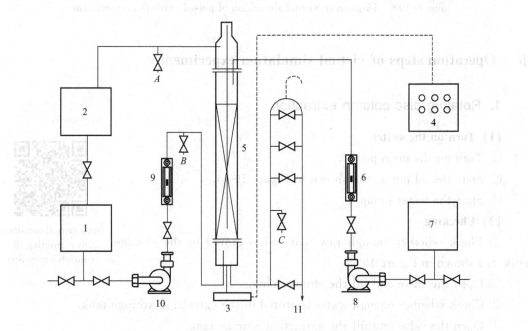

Figure 10-5　Pulsed extraction experiment diagram

1—Storage tank (benzoic acid-kerosene); 2—Collection tank (raffinate); 3—Pulsed system; 4—Control cabinet;
5—Packed (pulsed) column; 6,9—Rotameter; 7—Extractant storage tank (water); 8,10—Pump;
11—Discharge (extraction liquid) tube; A~C—Sampling point

Chapter 10　Extraction Experiment

diagram is shown in Figure 10-6.

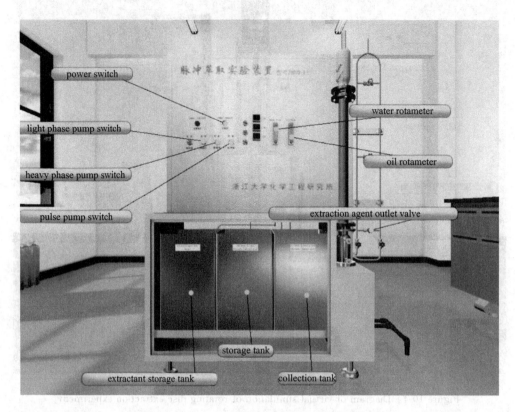

Figure 10-6 Diagram of virtual simulation of pulsed extraction experiment

Ⅳ Operation steps of virtual simulation experiment

1. Rotating disc column extraction

(1) Turn on the switch

① Turn on the main power.

② Start the oil pump (as shown in Figure 10-7).

③ Start the water pump.

Scan two-dimensional code: rotating disc column extraction

(2) Checking

① Check whether enough raw material is stored in the storage tank (as shown in Figure 10-8).

② Open the valve and fill the storage tank.

③ Check whether enough water is stored in the extractant storage tank.

④ Open the valve and fill the extractant storage tank.

(3) Experiment at the speed of 0

① Adjust the heavy phase (water) flowmeter so the flow rate is between 100~200mL/min (as shown in Figure 10-9).

Figure 10-7 Operation of control cabinet

Figure 10-8 Check whether the storage tank is full

Figure 10-9 Adjust the heavy phase (water) flowmeter

Figure 10-10 Interface between oil and water

② After the water flows in the extraction column for 5 minutes, adjust the light phase (kerosene) flowmeter so the flow rate is between 100~200mL/min.

③ Adjust the outlet valve of extraction phase until the interface between two phases in the quiet zone maintains constant (as shown in Figure 10-10).

④ The experiment should be stably operated for about 30 minutes. Under the condition of constant interface between two phases, open the sampling valve for sample analysis.

(4) Analysis of feed solution

① Move out of the tube, open the sampling valve of the feed solution and take the liquid at 2/3 of the tube (as shown in Figure 10-11).

② Take out the hydrometer and start from the smallest scale hydrometer to measure density first (as shown in Figure 10-12). Record the readings of the hydrometer.

③ Take 10mL feed solution out of the tube with the pipette and move it into the Erlenmeyer flask (as shown in Figure 10-13).

④ Drop two drops of indicator in the Erlenmeyer flask and shake the flask to fully mix the solution.

⑤ Put sodium hydroxide solution into the alkaline pipette.

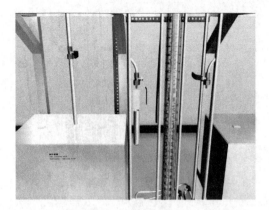
Figure 10-11 Take the feed solution

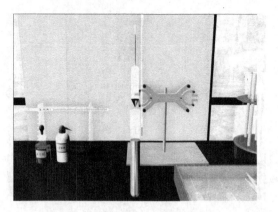
Figure 10-12 Use the hydrometer for measurement

⑥ Drop a small amount of sodium hydroxide into the Erlenmeyer flask and sufficiently mix the solution (as shown in Figure 10-14). Repeat the operation until the solution in the Erlenmeyer flask turns from yellow to green.

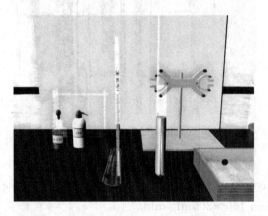
Figure 10-13 Use the pipette to transfer liquid

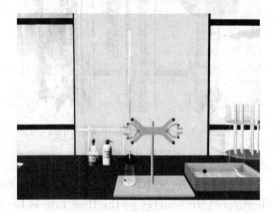
Figure 10-14 Titration

⑦ If the solution turns blue, it indicates that it contains excess sodium hydroxide and requires resampling analysis.

⑧ Record the amount of alkaline used for titration.

(5) Analysis of raffinate phase

① Move out of the tube. Open the sampling valve of raffinate. Take the liquid at 2/3 of the tube.

② The sample analysis is the same as that of feed solution by using neutralization titration method.

(6) Change the speed and repeat the experiment

① Speed switch (as shown in Figure 10-15) is used to adjust the speed.

② Keep the flow rate of water and kerosene unchanged. Observe the position of liquid at the column top.

③ Adjust the outlet valve of the heavy phase until the phase interface in the quiet zone maintains constant (as shown in Figure 10-16).

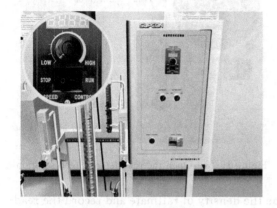
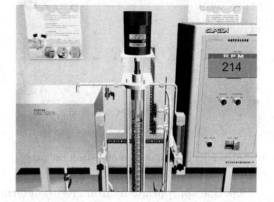

Figure 10-15 Speed switch   Figure 10-16 Adjust the outlet valve of heavy phase

④ The experiment should be stably operated for about 30 minutes.

⑤ Under the condition of constant interface between two phases, open the sampling valve for sample analysis.

⑥ The sample analysis is the same as that of feed solution by using neutralization titration method.

(7) Turn off the apparatus and clean the instruments

① Close the speed control valve and turn off the speed switch.

② Close the water flowmeter and kerosene flowmeter.

③ Turn off the oil pump switch, water pump switch and main switch.

2. Pulsed column extraction

(1) Check the apparatus and start the experiment

① Check whether there is sufficient solution in the extractant and feed storage tanks before the experiment. In case of insufficient solution, add solution into the tanks.

② Turn on the main power, heavy phase pump and light phase pump (as shown in Figure 10-17).

Scan two-dimensional code: pulsed column extraction

③ Open the rotameter of extraction agent. Add extraction agent into the whole column.

④ Open the rotameter of feed solution. Add the feed solution according to the amount of extraction agent (as shown in Figure 10-18).

(2) Make sampling analysis when pulsed extraction is not started

① Adjust the outlet valve of extraction liquid to maintain the flow stable. The interface between oil and water should be located between the outlet of extraction agent and raffinate.

② After the operation is stable for 30 minutes, open the sampling valve of raffinate to take samples to 2/3 of tube (as shown in Figure 10-19).

Chapter 10 Extraction Experiment

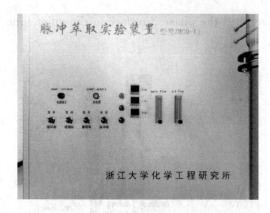
Figure 10-17　Operation of control cabinet

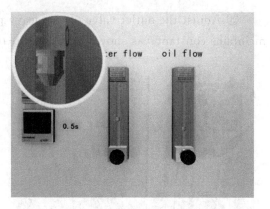
Figure 10-18　Open the rotameter of feed solution

③ Use appropriate hydrometer to determine the density of raffinate and record the reading of the hydrometer (as shown in Figure 10-20).

Figure 10-19　Raffinate sampling

Figure 10-20　Record the reading of hydrometer

④ Determine the composition of the raffinate with neutralization titration method.

⑤ Drop enough NaOH solution to the buret and take 10mL raffinate to the Erlenmeyer flask with pipette (as shown in Figure 10-21). Add indicator into the Erlenmeyer flask.

⑥ Make sure to make slow titration and completely mix the solution until the solution turns green (as shown in Figure 10-22).

⑦ Take another tube, open the sampling valve of feed liquid and take samples to 2/3 of the tube.

⑧ The method and steps of determining feed liquid are the same as those of determining raffinate.

⑨ After the determination, pour the residual feed liquid into the tank.

(3) Change the pulse and repeat the experiment

① Turn on the pulse pump and the switch to get pulses with a frequency of 0.5s (as shown in Figure 10-23).

② Keep the flow rate unchanged and readjust the two outlet valves at the top to locate the extraction interface between the outlets of extraction agent and raffinate (as shown in

Figure 10-24).

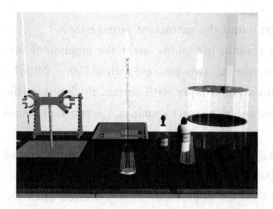

Figure 10-21 Use the pipette to transfer raffinate

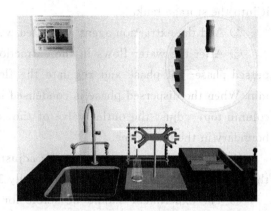

Figure 10-22 Titration

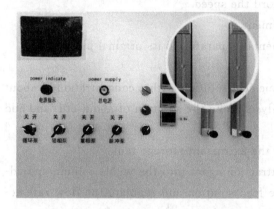

Figure 10-23 Turn on the pulsed pump switch

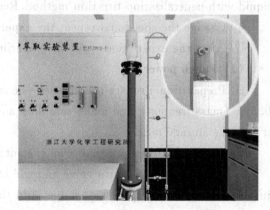

Figure 10-24 Interface between oil and water

③ Keep it stable for 30 minutes, measure and record the pulse amplitude. Determine the composition and density of raffinate.

④ Change the pulse frequency and repeat the experiment.

(4) Equipment arrangement after experiment

① Turn off the pulse switch and pulse pump switch.

② Turn off the heavy phase pump switch and light phase pump switch.

③ Close the two flowmeters.

④ Close the outlet valve of extraction liquid.

⑤ Turn off the main power switch.

Ⅴ Experimental method and procedure

1. Operating procedures

(1) Rotating disc column

① Preparation. Prepare the kerosene-benzoic acid solution with a concentration of about

0.3% (mass fraction) and make sampling analysis. Take the solution of about 50L and add it into the storage tank.

② Add the extraction agent (distilled water) into the extractant storage tank.

③ After the water flows in the extraction column for 5min, start the pipeline of dispersed phase—oil phase and regulate the flow rate of two phases within 100～200mL/min. When the dispersed phase is condensed into a liquid layer with certain thickness at the column top, adjust the outlet valve of the continuous phase to maintain a constant phase boundary in the quiet zone.

④ Control the external energy by adjusting the speed. Gradually increase the speed of the operation. The speed should be generally 100～700r/min.

⑤ The experiment should be stable for about 30 minutes every time. Then open the sampling valve to analyze the samples. Determine the composition of raffinate and extraction liquid with neutralization titration method. Record the speed.

⑥ Change the speed and repeat the experiment.

⑦ After the experiment, put the experimental apparatus to its original places.

(2) Pulsed power

① Preparation. Prepare the kerosene-benzoic acid solution with a concentration of about 0.3% (mass fraction) and make sampling analysis. Take the solution of about 50L and add it into the storage tank.

② Add the extraction agent (water) into the extractant storage tank.

③ Open the flow control valve and add extraction agent into the whole column and adjust the flow rate to the desired value. Add the feed solution and regulate the flow rate according to the amount of extraction agent. In the experiment, keep the flow rate constant and adjust the outlet valve of extraction liquid to keep the interface of oil and water between the outlet of extraction agent and raffinate (as shown in Figure 10-10).

④ Turn on the pulse steering motor and adjust the motor frequency to the desired value.

⑤ The experiment operation should be stable for about 30 minutes every time. Then open the sampling valve to take the samples. Determine the composition of raffinate and extraction liquid with neutralization titration method. Record the pulse parameters.

⑥ Change the pulse parameters and repeat the experiment.

⑦ After the experiment, arrange the experimental apparatus to its original places.

2. Attention

① Be careful when the speed is adjusted and gradually increase the speed. Do not abruptly increase the speed. From the perspective of fluid mechanic performance, if the speed is too high, flooding would occur resulting in unstable operation.

② Throughout the experiment, the interface of two phases at column top must be properly located between the outlet of light phase and inlet of the heavy phase and should be stable.

③ As disperse phase and the continuous phase excessively remain at column top and bottom, when the operating conditions are changed, the operation must be stable for a long time, or the errors would be significant.

④ The actual volume tric flow rate of kerosene is not necessarily the reading of the flowmeter. When the values of actual flow rate of kerosene are required to be used, they can be used by correcting the readings of the flowmeter with flow rate correction formula.

Ⅵ Experimental report

1. Calculate the number of mass transfer units N_{OE} under different speed/pulse.
2. Calculate the height of a mass transfer unit H_{OE} under different speed/pulse.
3. Calculate the volumetric mass transfer coefficient $K_Y a$ under different speed/pulse.
4. Analyze the experimental results according to the theory of mass transfer.

Ⅶ Questions

1. Analyze and compare the similarities between extraction column and absorption and distillation column.
2. How is external energy adjusted and measured with the extraction column in the experiment?
3. Analyze the impact of the changes in rotary speed or pulse parameters on mass transfer coefficient in extraction and extraction efficiency based on the experimental results.
4. Which methods can be used to determine the composition of feed, extraction phase and raffinate phase?
5. Does flooding occur in extraction and how to determine the flooding?

Chapter 11 Experiment for the Determination of Drying Characteristic Curve in Tunnel Dryer

I Experimental purposes and requirements

1. To understand the structure, flow and operational methods of tunnel dryer.

2. To plot the drying characteristic curves ($X \sim \tau$, $U \sim X$) of material in the constant drying condition and calculate the critical moisture content X_c, equilibrium moisture content X^* and drying rate U_{cs} in the period of constant velocity.

3. To calculate the mass transfer coefficient k_W and heat transfer coefficient α at constant velocity.

4. Change operating conditions such as temperature or gas velocity, fit drying characteristic curve under different air parameters, determine critical moisture content, equilibrium moisture content and drying rate, mass transfer coefficient and heat transfer coefficient in constant rate stage.

II Experiment principle

Different moistures contained in the material will surely cause significant changes in the process of drying. In order to reduce the impact, the wet material is dried in the constant drying conditions (i.e. the temperature, humidity and velocity of air as drying medium and its contact with the material are maintained constant) in the experiment. The relationship between the drying time τ and the moisture content X of the material can be obtained by measuring the weight loss of the wet material at different time, and the drying rate curve $X \sim \tau$ of the material can be obtained by sorting out the data. Through the weighing of wet material at different time, the relationship between drying time τ and weight G of wet material can be obtained. The drying curve $X \sim \tau$ and drying rate curve $U \sim X$ can be obtained by processing the data.

By studying drying rate curve, the drying process mainly includes the constant rate and falling rate drying periods.

Constant rate drying period. Wet material surface is completely wetted by unbound water. During the period of moisture evaporated from the surface of material, the rate of moisture in the wet material diffusing to the surface is equal to or greater than the rate of moisture evaporated from the surface. The surface of the material remains wet and the surface temperature is the wet bulb temperature t_W in the air.

Falling rate drying period. The rate of the moisture in the wet material diffusing to the surface is lower than the rate of moisture evaporated from the surface. The temperature of material increases or the surface becomes dry. The drying enters the falling rate period. As the material continues to be dried, the moisture contained in it is less, so the rate of the moisture in the material diffusing to the surface is lower and the drying rate is also reduced until the moisture content of the material is equal to the equilibrium moisture content X^* in the air. The drying rate in falling rate period is determined by the structure, shape and size of dry material, and it has no significant relation with the status of the drying medium, therefore the falling rate period is also called the control stage of material internal migration.

Critical moisture content X_c is the content dividing constant rate drying and falling rate drying. Critical moisture content is critical to the drying mechanism and dryer design.

1. Determination of drying characteristic curve ($X \sim \tau$, $U \sim X$)

Drying rate is the rate to remove the moisture from per unit of surface area in unit time through drying, represented by U, so

$$U = \frac{dW}{A\,d\tau} = -\frac{G_c}{A} \times \frac{dX}{d\tau} \tag{11-1}$$

where U —drying rate, kg/(m² · s);

  A —drying area, m²;

  G_c —material oven dry weight, kg;

  X —material dry basis moisture content, kg water/kg dry air.

According to the relationship curve $G_i \sim \tau$ between the material weight and time measured by computer and weight sensor at different time, the dry basis moisture content X_i of the material at different time can be obtained

$$X_i = \frac{G_i - G_c}{G_c} \tag{11-2}$$

where X_i —moisture content of material at τ_i, kg water/kg dry air;

  G_i —material weight (including accessories weight) at τ_i, kg;

  G_c —material oven dry weight (including accessories weight), kg.

According to equation (11-2), the values of X_i corresponding to the time τ_i can be obtained. Based on the values, the drying curve $X \sim \tau$ can be drawn. Find X_c on the curve $X \sim \tau$ and then take typical points to draw graphs on the curve $X \sim \tau$ to obtain the slope $\frac{dX}{d\tau}$. Then calculate the drying rate U_i according to equation (11-1) and draw the drying rate curve $U \sim X$. Find X^* and U_{cs} in the figure $U \sim X$.

2. Determination of mass transfer coefficient and heat transfer coefficient in the condition of constant rate drying

If the water volume passed from material surface per unit area to air in unit time is G_W,

Then
$$G_W = k_W(H_W - H) \quad (11\text{-}3)$$

According to the mass transfer rate equation, the constant rate in drying stage U_{cs}:
$$U_{cs} = G_W = k_W(H_W - H) \quad (11\text{-}4)$$
So
$$k_W = \frac{U_{cs}}{H_W - H} \quad (11\text{-}5)$$

If the heat transferred from air to material surface per unit area in unit time is Q, then
$$Q = \alpha(t - t_W) \quad (11\text{-}6)$$

Based on the energy balance relation, this heat should be equivalent to the energy consumed by vaporization.
$$Q = r_W U_{cs} = r_W k_W (H_W - H) \quad (11\text{-}7)$$
Thus
$$\alpha(t - t_W) = r_W U_{cs} = r_W k_W (H_W - H) \quad (11\text{-}8)$$
i. e.
$$\alpha = \frac{r_W U_{cs}}{t - t_W} = \frac{r_W k_W (H_W - H)}{t - t_W} \quad (11\text{-}9)$$

where U_{cs} —drying rate in the period of constant rate drying, $kg/(m^2 \cdot s)$;

Q —heat transferred from air to material surface per unit area in unit time, W/m^2;

G_W —water volume passed from material surface per unit area to air, $kg/(m^2 \cdot s)$;

k_W —mass transfer coefficient in the period of constant rate drying, $kg/(m^2 \cdot s)$;

H —air humidity, kg water/kg dry air;

H_W —saturated humidity of air, kg water/kg dry air;

α —convective heat transfer coefficient between material surface and air in the period of constant rate drying, $W/(m^2 \cdot \text{°C})$;

t_W —wet bulb temperature of air in dryer, °C;

t —dry bulb temperature of air in dryer, °C;

r_W —latent heat of water under t_W, J/kg.

In this experiment, the relationship $G_i \sim \tau$ between material weight loss and time is measured, that is, the curves $X \sim \tau$, $U \sim X$ are obtained, and then X_c, X^* and U_{cs} are calculated. Finally, the mass transfer coefficient k_W and heat transfer coefficient α at constant rate stage are achieved.

Ⅲ Experimental apparatus and process

The experimental apparatus is shown in Figure 11-1 and the virtual simulation is shown in Figure 11-2. The fan blows air into the preheat chamber for preheating. Cold air is electrically heated to T_1 and flows into the dryer to heat the material and then is directly discharged into the air.

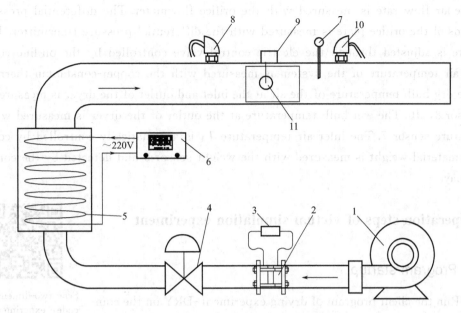

Figure 11-1 Drying process diagram

1—Fan; 2—Orifice flowmeter; 3—Differential pressure sensor; 4—Electric control valve; 5—Heater;
6—Temperature control system; 7—Wet bulb temperature sensor;
8,10—Temperature sensor; 9—Weight sensor; 11—Dry material

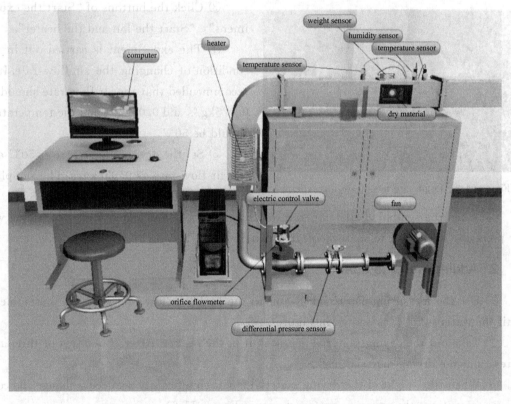

Figure 11-2 Diagram of virtual simulation drying experiment

Chapter 11 Experiment for the Determination of Drying Characteristic Curve in Tunnel Dryer

The air flow rate is measured with the orifice flowmeter. The differential pressure at both ends of the orifice plate is measured with the differential pressure transmitter. The air flow rate is adjusted through the electric control valve controlled by the on-line computer. The air temperature of the system is measured with the copper-constantan thermocouple. The dry bulb temperature of the air at the inlet and outlet of the dryer is measured with the sensor 8, 10. The wet bulb temperature at the outlet of the dryer is measured with the temperature sensor 7. The inlet air temperature T_1 is automatically controlled by computer. The material weight is measured with the weight sensor 9 and detected by the computer for display.

IV Operation steps of virtual simulation experiment

1. Program startup

① Run the client program of drying experiment -DRY on the computer desktop. Enter the control interface of drying experiment (as shown in Figure 11-3).

Scan two-dimensional code: experiment for the determination of drying characteristic curve in tunnel dryer

② Click on the "login" button, enter the account number and password. Click to enter (as shown in Figure 11-4).

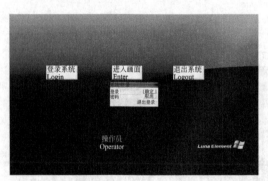

Figure 11-3 Login interface of drying experiment

③ Click the buttons of "Start the experiment", "Start the fan and the heater".

④ The experiment is carried out in the condition of changing the air flow rate. It is recommended that the air flow rate should be 0.025kg/s and 0.06kg/s and the temperature should be 50℃.

⑤ Set the dryer temperature as 50℃ and the air flow rate as 0.025kg/s. The sampling time interval should be 60s.

⑥ When the symbol of air temperature is displayed as a line and a prompt of adding water pops up, add water to soak the drying material.

2. Adding water to soak

① Use the measuring cup to add water to the beaker below the wet bulb thermometer until the water is full (as shown in Figure 11-5).

② Take out the drying material and soak it in the beaker. After about two or three minutes, put the drying material back.

③ When the weight of the drying material is not significantly reduced, change the conditions to continue the experiment (as shown in Figure 11-6).

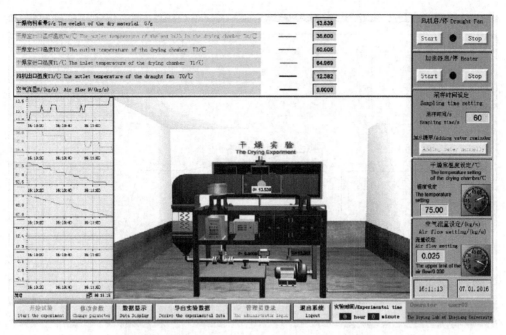

Figure 11-4 Control interface of drying experiment

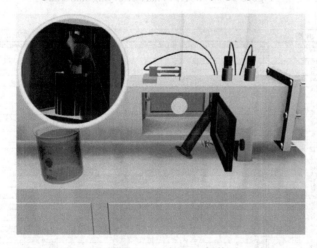

Figure 11-5 Add water to the wet-bulb thermometer

3. Parameter changing for repeating the experiment

① Increase the air flow rate to 0.06kg/s and keep on the experiment.

② Keep on observing the experiment until the weight of the drying material is not significantly reduced.

③ Save the experimental data.

4. Data storage

① Click the "Export Data" on the screen (as shown in Figure 11-7). Click on the export icon on the top left of the screen.

② Select all experiments and click the "OK" button. Export the data and save them on

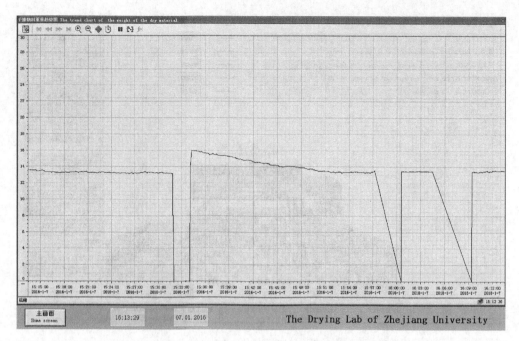

Figure 11-6　Weight chart of drying material

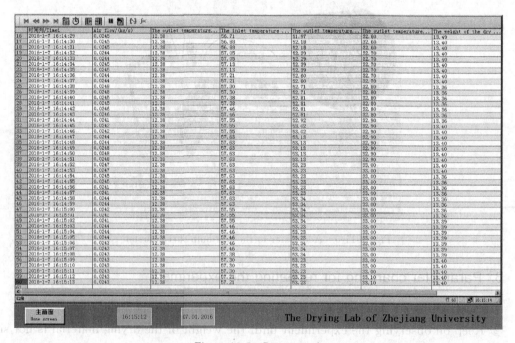

Figure 11-7　Data record

the desktop.

5. Program close

① Click on the "main screen" to return to the interface of operation.
② Turn off the heater.
③ Power off the fan when the dry bulb temperature drops to room temperature.

④ Click to exit the system.

V Experimental method and procedure

1. Operating procedures

① Turn on the computer. Run the client program of drying experiment. Enter the control interface of drying experiment.

② Click the buttons of "Power switch" and "Start the experiment". Start the fan. Open the electric control valve to its full open position to set the maximum flow rate of air. Start the heater and set the inlet air temperature to a suitable value (60~90℃). When the symbol of air temperature is displayed as a straight line on the computer screen, add water to soak the drying material. Meanwhile, add water to the beaker below the wet bulb thermometer. Then the drying experiment automatically starts. When the weight of material does not change significantly, it indicates that the drying is completed under this operating condition.

③ Change the air flow rate or temperature, repeat the experiment.

④ After the test, save the data.

⑤ Turn off the power of the heater. When the dry bulb temperature drops below 50℃, turn off the fan power and main power, exit the system.

2. Attention

① Carefully use the weighing sensor as it can be easily damaged. Do not weigh heavy objects.

② Be sure to start the fan first and then the heater.

VI Experimental report

1. Plot the drying curve graphs and drying rate curves.
2. Calculate the critical moisture content of materials.
3. Calculate the convective heat transfer coefficient between material surface and air in the period of constant rate drying.

VII Questions

1. How to acquire the constant drying condition from the experiment?
2. How to calculate the air humidity H and H_W from the experiment?
3. What are the impacts of the change the air velocity or temperature in critical moisture content and equilibrium moisture content?
4. What advantages and disadvantages are the air circulation dryer and exhaust emission dryer?

参 考 文 献

[1] 浙江大学化学过程实验室.过程工程原理实验.杭州：浙江大学，2008.
[2] 北京大学，南京大学，南开大学.化工基础实验.北京：北京大学出版社，2004.
[3] 陈寅生.化工原理实验及仿真.第2版.上海：东华大学出版社，2008.
[4] 王存文.化工原理实验（双语）.北京：化学工业出版社，2014.
[5] 田维亮.化工原理实验及单元仿真.北京：化学工业出版社，2015.
[6] 吴嘉.化工原理仿真实验.北京：化学工业出版社，2001.